Acknowledgement

This publication has been made possible with the support of the Army History Research Grants Scheme; and the support of the National Executive, State Branches and members of the Defence Reserves Association.

Published by Echo Books.

Echo Books is an imprint of Superscript Publishing Pty Ltd, ABN 76 644 812 395

Registered Office: PO Box 669, Woodend, Victoria, Victoria, 3442.

www.echobooks.com.au

Copyright ©Defence Reserves Association

Creator: Andrew J Kilsby, Author

Title: The Reservists, A History of the Defence Reserves Association 1970-2020

ISBN: 978-1-922603-17-3

 A catalogue record for this book is available from the National Library of Australia

Book layout and design by Peter Gamble, Canberra
Set in Garamond Premier Pro Display, 12/17, and MinervaModern.

The Reservists

A History of the Defence Reserves Association 1970-2020

Andrew J Kilsby

With a foreword by
His Excellency the Honourable General David J. Hurley,
AC, DSC (Retd), Governor General of Australia

Dedicated to
Major-General Paul Alfred Cullen AC CBE DSO* ED FCA
1909-2010

(Portrait by Mathew Lynn, 2002, National Portrait Gallery, Canberra)

Contents

About the author	vii
Foreword	ix
Introduction	xi
Abbreviations	xiii

Part One: The CMF Association 1970–1987

1. The CMF at the Crossroads	1
2. Wax and Wane	15
3. Towards a National Association	31
4. The Total Army and the Dibb Report	47

Part Two: The Army Reserve Association, 1987–1992

5. Interregnum	61
6. Future Directions	77

Part Three: The Defence Reserves Association, 1993–2020

7. At the Crossroads–Again	97
8. A21 Alarms	111
9. Another New Era	135
10. Raise, Train and Sustain	155
11. Plans BEERSHEBA and SUAKIN	179
12. Disappointment and Optimism in Equal Measure	201
Epilogue–The Sum of the Parts	221

Annexes

Annex A: CMF Association, 1970-1986	227
Annex B: Army Reserve Association, 1987-1992	230
Annex C: Defence Reserves Association, 1993-2020	232
Select Bibliography	237
Notes	241
Index	273

About the Author

Dr Andrew Kilsby is an independent historian and published author. He has had careers in military and diplomatic service both in Australia and Asia, in public relations & corporate communications. His history interests span military, business, and biographical history. He is a graduate of RMC Duntroon and holds a PhD in history from UNSW.

A co-founder of Military History & Heritage Victoria Inc. Dr Kilsby has published several histories, convened and presented at military history conferences, arranged exhibitions and written articles and biographies. Dr Kilsby has many *Citizen Soldier* family connections, from colonial militia through both World Wars and most recently with a family member serving with 4th Combat Engineer Regiment.

Military related titles by Andrew Kilsby include:

The Case of Eichengruen-Edwards and Continental Tyres

The Forgotten Cruiser: HMAS *Melbourne 1913-1928*

Before the Anzac Dawn – contributed chapter.

The Riflemen: A History of the NRAA 1888-1988

In the Field and On the Field: A history of Australian Army Rugby Union

fallen leaves: the Oakleigh WWI Honour Rolls

The Bisley Boys

Lions of the Day

Foreword

His Excellency General the Honourable David Hurley AC DSC (Retd)
Governor-General of the Commonwealth of Australia

The recounting of the history of the Defence Reserves Association as presented in this important book leads to the thought that just as beauty is in the eye of the beholder so it seems are views on the importance, relevance and roles of the Reserves. Having served in the Army for forty of the DRA's fifty-year history and been both an observer and decision maker in the debate on the role and tasks of the Reserves, I found the author's accurate recording and analysis of events compelling reading.

The debate concerning the relevance, responsibilities and roles of the Reserves or part-time forces in the Australian Defence Force is of great significance to Australia's national security, yet it is matter little understood and pursued outside a small group of advocates and commentators. Since 1970 the platform for advocates for the Reserves has been the Defence Reserves Association. This book is a faithful retelling of the Association's work and advocacy.

I am confident that readers of this book will have similar reactions to mine: recalling eminent characters of the past; reliving events and decisions of great significance to the future of the Reserves; and gleaning information that provoked an 'Ah, so that's why' response. I can recall some examples from my own experiences.

In 1979, as Adjutant of Sydney University Regiment, I assisted in the planning and conduct of the annual regimental dinner held in the Great Hall of the University of Sydney. In attendance, apart from many senior Reserve officers, were the Chief of the General Staff (CGS), Field Force Commander, Commander Training Command and Commander Logistics Command. In his after dinner speech the CGS stated that 'this was the most important regimental dinner held in the Army'. As a young officer I was left with no doubt at the importance of the Reserves! Yet as Andrew informs us, the import of this statement was not necessarily evident in decisions concerning the employment of the Reserves during the 1980s.

Fast forward to February 1991, when as SO1 OPS Headquarters 2nd Division, I was present when Major General Paul Cullen delivered the Blamey Oration in Sydney. As discussed in this book, this was a seminal speech that captured the history of the competing interests and perspectives between Regular and Reserve elements of the Army on the role of the Reserves and addressed Cullen's views on the force structure implications of the 1987 White Paper. The speech was a very clear statement of the DRA's hopes and ambitions for the Reserves. I detected from a number of the questions that night that there was a split between the views of the younger and older generations of Reservists that would need to be resolved if the DRA was to speak for all Reservists.

This history of the Association rightly pays tribute to Major General Cullen for his leadership and tenacity in arguing the Association's position on the utilisation of the Reserves and its conditions of service. As the book recounts, General Cullen continued in an active role well after he retired from office in the Association and much that the Association achieved rests at his feet.

The period 2009 to 2014 was a time of considerable turbulence for the Reserves but, importantly, the DRA experienced a significant improvement in its access and participation in decisions affecting the Reserves. During this period I served as both VCDF – with control of the Cadets, Reserves and Employer Support Division – and as CDF. It was a period in which the ADF as a whole was buffeted by major structural changes as a consequence of the Strategic Reform Program and enormous pressures on the Defence budget. Despite this, and with the support and input of the DRA, progress with made with two projects that would greatly assist in implementing positive change for the Reserves – Plans Beersheba and Suakin. As recorded in this book, the two plans met 80% of the DRA's ambitions for the Reserves. The DRA should be congratulated for its positive role in the development and implementation of both plans.

Andrew Kilsby, in this extraordinarily well-researched book, does justice to the perseverance and achievements of those who led and served the Association at the national and state levels. From 'The CMF at the Crossroads', which starts the post-Vietnam War period from 1970 to 1987, to 'Disappointment and Optimism in Equal Measure', which considers decisions made in 2020, Andrew's recounting of the history of the Association and its engagement with Government and Defence captures in detail the frustrations, small and major triumphs, disappointments, and policy coups of the past fifty years. It is true to say that this book illustrates that the Association has been 'patient if nothing else' as it pursued its goals.

Introduction

Major-General Jim Barry, AM MBE RFD ED (Rtd)

It gives me considerable pleasure to provide an introduction to this history of the Defence Reserves Association. It is especially gratifying to me to have been designated as the project manager for the publication by the DRA Executive Committee, under the leadership of former president Major-General Paul Irving, AM PSM RFD (Retd). I would like to thank all of the members of the DRA involved in supporting this project and for their help in bringing the history project to a successful conclusion.

The history is dedicated to Major-General Paul Alfred Cullen AC, CBE, DSO & Bar, ED (13 February 1909 – 7 October 2007), who founded the original CMF Association and remained active in its evolution through to the DRA. Without doubt, it is a fitting dedication, for without his vision, drive and determined advocacy for Reserves, they may well have been emasculated many years ago to the detriment of community support for the Defence Force and its effectiveness today. Cullen was supported in his efforts by a range of capable and experienced Reservists throughout the long journey of the Association.

I would also like to thank the Governor-General, David Hurley, for his kind foreword. As a former Chief of Defence Force, he has been aware of the many issues that have been raised on behalf of Reservists by the DRA and has been actively involved in Reserve matters over many years.

The historian, Dr Andrew Kilsby, has done a thorough and commendable job of uncovering and describing both the minutiae and the grand debates of the DRA since its inception more than 50 years ago. This history has reminded me, despite my own many years in the association, of how little sometimes we know of the foundation efforts of our predecessors, let alone their struggles and efforts to overcome the prejudices of permanent defence (both uniformed and civilian) and Government members alike, towards reserve service. To a very large extent, these issues are behind us, with Reserves incorporated into the modern Defence Force like never before.

Histories of unit and other associations in the Defence Force, like the DRA, are important, not so much because of the conduct of the associations themselves or even the detail of the very many committee and association activities over many years, but because they support a richer knowledge of the context in which they exist. This history does that. It contains a wealth of insights into the workings of Defence, the Reserve forces themselves and the evolution of policy affecting the Reserves. Like the DRA itself, this history is more than the sum of its parts.

Abbreviations

ABC Australian Broadcast Commission

AC Companion of the Order of Australia

ACA-R Assistant Chief Army Reserves

ACDF-R Assistant Chief Defence Force Reserves

ACGS-R Assistant Chief General Staff Reserves

ACM Air Chief Marshal

ACRES-A Assistant Chief Reserves – Army

ACTU Australian Council of Trade Unions

ADA Australian Defence Association

ADC Aide-de-Camp

ADF Australian Defence Force

ADFA Australian Defence Force Academy

ADSO Alliance of Defence Service Organisations

AF Air Force

AG Advocate-General

AGM Annual General Meeting

AIC Australian Instructional Corps

AIF Australian Imperial Force

AK Knight, Order of Australia

ALP Australian Labor Party

AM Member, Order of Australia

ANR Australian Naval Reserve
AO Officer of the Order of Australia
ARA Australian Regular Army
ARAC Army Reserve Advisory Council
AResA Army Reserve Association
ARes/ARES/AAR Army Reserve
ArFFA Armed Forces Federation of Australia
ARRC Army Reserve Review Committee
ASPI Australian Strategic Policy Institute
AVADSC Australian Veterans and Defence Services Council
A21 Army into the Twenty First Century

BRIG Brigadier

CA Chief of Army
CAPT Captain
CAS Chief of Air Staff
CB Companion of the Order of the Bath
CBE Commander of the Most Excellent Order of the British Empire
CCR Community Consultation Report
CD Canadian Forces Decoration
CDF/CDFS Chief of Defence Force/ Chief of Defence Force Staff
CESAR Committee of Employer Support Army Reserve
CESRF Committee for Employer Support of Reserve Forces
CFTS Continuous Full Time Service
CGS Chief of the General Staff
CGSAC Chief of the General Staff Advisory Committee
CIMIC Civil Military Cooperation
CIT Common Induction Training
CMDR Commander

CMFA Citizen Military Forces Association

CMF Citizen Military Forces

CMG Companion of the Most Distinguished Order of St Michael and St George

CO Commanding Officer

COL Colonel

COVID-19 Coronavirus 2019

CRES Chief of Reserves

CRESD Cadet, Reserve and Employer Support Division

CRT Common Recruit Training

CSC Conspicuous Service Cross

CSM Conspicuous Service Medal

CVO Companion of the Royal Victorian Order

DCC Defence Consultative Committee

DCM Distinguished Service Medal

DCRES Deputy Chief of Reserves

DFC Distinguished Flying Cross

DER Defence Efficiency Review

DFRDB Defence Forces Retirement Benefits Fund

DFRT Defence Force Remuneration Tribunal

DFWA Defence Force Welfare Association

DGANCR Director General Australian Navy Cadets and Reserves

DGR-A Director General Reserves-Army

DGR-AF Director General Reserves-Air Force

DGR-N Director General Reserves-Navy

DHOAS Defence Housing Ownership Assistance Scheme

DNR Director Navy Reserves

DOD Department of Defence

DRA Defence Reserves Association

DRES-AF Director of Reserves-Air Force

DRP Defence Reform Program

DRSC Defence Reserves Support Committee
DRSD Defence Reserves Support Day
DSC Distinguished Service Cross
DSM Distinguished Service Medal
DSO Distinguished Service Order
DSO* Distinguished Service Order and bar
DVA Department of Veterans Affairs
DWP Defence White Paper

ED Efficiency Decoration
ESO Ex-Service Association
ESORT Ex-Service Organisations Round Table
ESPS Employer Support Payment Scheme
FLTLT Flight Lieutenant
fnu full name unknown
FSR Force Structure Review
F/T/FTS Full Time/Full Time Service

GDP Gross Domestic Product
GFC Global Financial Crisis
GOC General Officer Commanding
GPCAPT Group Captain
GRES General Reserve

HMSP-A Head Modernisation and Planning Army
HQ ADF Headquarters Australian Defence Force
HRP Head of Reserve Policy
HRR High Readiness Reserves

IOOF Independent Order of Odd Fellows

ISO Companion of the Imperial Service Order

IT Information Technology

JSCFADT Joint Standing Committee on Foreign Affairs, Defence and Trade

KBE Knight Commander of the Most Excellent Order of the British Empire

KCMG Knight Commander of the Distinguished Order of St Michael and St George

KCVO Knight Commander of the Royal Victorian Order

KOC Kindred Organisations Committee

KPMG Klynveld Peat Marwick Goerdeler

KStJ Knight of the Order of St John

KtON Knight of the Order of Orange-Nassau

LLB Bachelor of Laws

LS Long Service

LT Lieutenant

LTCDR Lieutenant Commander

LTCOL Lieutenant Colonel

LTGEN Lieutenant General

LVO Lieutenant of the Royal Victorian Order

MAJ Major

MAJGEN Major General

MBE Member of the Most Excellent Order of the British Empire

MC Military Cross

MC* Military Cross and bar

MCRA Military Compensation and Remuneration Act

MCS Military Compensation Scheme

MD Medical Doctor

MG Medal of Gallantry

MHR Member of the House of Representatives
MINDEF Minister for Defence
MLOC Minimum Level of Operational Capability
MONUR Monash University Regiment
MP Member of Parliament
MRCA Military Remuneration and Compensation
MStJ Member of the Order of St John
MUR Melbourne University Regiment
MVO Member of the Royal Victorian Order

NCF National Consultative Framework
NCO Non-Commissioned Officer
NORFORCE North-West Mobile Force
npn no page number
NSAA National Servicemen's Association of Australia
NSW New South Wales
NT Northern Territory

O St J Officer of the Order of St John
OA/OAM Member, Order of Australia
OBE Officer of the Most Excellent Order of the British Empire
OLOC Operational Level of Capability
OMM Order of Military Merit (Canada)
ORBAT Order of Battle

PAF Permanent Air Force
PM Prime Minister
PNF Permanent Naval Force
PSM Public Service Medal
P/T/PTS Part Time/Part Time Service

QC Queen's Counsel

RAA Royal Australian Artillery
RAAC Royal Australian Armoured Corps
RAAF Royal Australian Air Force
RAAMC Royal Australian Army Medical Corps
RAMSI Regional Assistance Mission to Solomon Islands
RAN Royal Australian Navy
RANR Royal Australian Navy Reserve
RFD Reserve Forces Decoration
RFD Reserve Forces Day
RFSD Reserve Forces Support Day
RFSU Regional Force Surveillance Units
RMC Royal Military College Duntroon
RMW Reserve Modernisation Workshops
RNSWR Royal New South Wales Regiment
RRES Ready Reserve
RSD Reserve Service Days
RSL Returned and Services League
RSM Regimental Sergeant Major
RTA Restructuring the Army
RUSI Royal United Services Institute
RVR Royal Victorian Regiment

SA South Australia
SER Self-Employed Reservists
SERCAT Service Categories
SQNLDR Squadron Leader
SRCA Safety Rehabilitation & Compensation Act
SRP Strategic Reform Program

SVN South Vietnam

TWM Total Workforce Model

UK United Kingdom
UN United Nations
UNSW University of New South Wales
USA United States of America
USAF United States Air Force
USMC United States Marine Corps

VC Victoria Cross
VCDF Vice Chief Defence Force
VD Volunteer Decoration
VEA Veterans Entitlements Act
VP Vice-President

WA Western Australia
WWI World War I
WWII World War Two

2 MD 2nd Military District
5 MD 5th Military District

Part One
The CMF Association
1970–1987

1.
The CMF at the Crossroads

The end of the 1960s and the beginning of the 1970s saw the concept of Citizen/ Reserve service increasingly challenged. Circumstances were such that many felt not only within the command structure but also in the Parliaments and community that the concept of Citizen/Reserve service needed defending.[1]

The Citizen Military Forces Association (CMFA) was created by Major General Paul Cullen CBE DSO ED, who had been the CMF Member of the Military Board to 1966. The primary reason was to help the CMF arrest the fall in political influence and connections felt through declining numbers of parliamentarians who had CMF (or even regular) service.[2] The first formal meeting of those interested in the concept of a CMFA was held on 18 March 1970 at the Imperial Services Club in Sydney followed by a further meeting on 9 April when a constitution and office bearers were agreed to.[3] The April meeting was attended by Cullen, Major General Allan C. Murchison MC ED, Brigadier Sir John E. Pagan CMG MBE ED and three others.[4]

The Imperial Services Club for Australian military officers returning from war service, was formed in 1917. As Reserves chronicler Dayton McCarthy wrote in 2002, 'the Sydney-based *Citizen Soldier* lobby was and still is a formidable force because traditionally Eastern Command and 2nd Division have been the largest command and CMF division respectively... For a long time, the colourful term 'Rum Corps' has been used to describe this Sydney CMF/ Reserve 'establishment'.[5] The Imperial Services Club was where the 'establishment' met.[6]

When the CMFA was formed, it was only natural that this was where the first and subsequent meetings were held. Paul Cullen was the driving force but on paper at least he was backed by a well-connected and influential power group of councillors. At the beginning however,

Cullen and his secretary, Lieutenant Colonel John R. Dart OBE RFD ED at the time the commanding officer of the Sydney University Regiment, 'virtually ran the CMFA by themselves.'[7]

By August 1970 Cullen was sending out application forms to interested individuals—'former and serving officers of all services'—to join the new Association. In the covering letter he wrote:

> This Association realizes that unless the regular officers serving with the CMF are satisfied, the CMF efficiency will be impaired. Therefore, we are anxious to improve the rates of pay and conditions of all serving officers of the Services, particularly in regard to aspects of housing and retirement benefits.[8]

This overt attempt to attract both regular and reserve officers to the cause initially fell on largely deaf ears. In 1970 only three men were noted as members of the 'National Council' of the CMFA: Cullen himself, the Treasurer Lieutenant Colonel Douglas Pennicook ED and Secretary John Dart. As Victoria was the next State (Southern Command) with 3 Division and numerous other units, on 23 April Dart wrote to Major General Stuart M. McDonald CBE MC ED, the former commander 3 Division, asking him to form a State Committee, keep the National Council informed and enclosed Minutes, a constitution and six membership application forms (subsequently McDonald asked for 10 sets).[9] Yet it would be another seven years before a branch was formed in Victoria.

By 1972, momentum in the development of the association had grown. The National Council, listed in the inaugural edition of *Citizen Soldier*, the CMFA magazine, in January 1972 now consisted of:

President: Major General Paul Cullen
Executive Director: Major Gilbert E. Cory MC DCM
Treasurer: Colonel D. Pennicook

Councillors:
Major General John A. Bishop DSO OBE ED (NSW)
Brigadier Sir John Pagan CMG MBE ED (NSW)
Brigadier Richard ('Dick') T. Eason MC ED (Vic)
Brigadier James ('Jim') H. Thyer CBE DSO (SA)
Brigadier Jack L. Amies CBE ED (Qld)
Brigadiers John B. Roberts MBE ED (WA)
Major J. R. W. Crook
Editor *Citizen Soldier*: Lieutenant Colonel Anthony ('Tony') C. Browne, ED
Publisher *Citizen Soldier*: Captain Brian Nebenzahl

Also attending these meetings was John Dart, the initial secretary and who would later become CMFA treasurer, and Major General Murchison, who, like Cullen, would become the CMF Member on the Military Board and in turn, President of the CMFA.[10] Almost without exception those officers had seen operational service in World War II either as Militia or in the Australian Imperial Force (AIF).

In addition, the following distinguished officers were offered Life Membership—all but Herring were NSW based:

> Lieutenant General Sir Edmund F. Herring KCMG KBE CBE DSO MC ED
> Major General Sir Denzil Macarthur-Onslow Kt CBE DSO
> Major General Sir Ivan Dougherty Kt CBE DSO* ED
> Major General Kenneth W. Eather CB CBE DSO ED
> Major General Sir Victor Windeyer KBE CB DSO ED
> Major General Noel W. Simpson CB CBE DSO* ED

In later years Life Member invitations were also extended to:

> Major General Sir Robert J. H. Risson CB CBE DSO O St J ED (Vic)
> Brigadier Sir Thomas Eastick CMG DSO ED
> Brigadier Sir Bernard Evans DSO ED (Vic)[11]

The CMFA constitution was surprisingly open, not just to members of the CMF past and present, but also to 'members and ex-members of the Australian Navy, Army and Air Force both Regular, Citizen and Reserve.'[12] In addition, 'Members of the public both male and female, Corporations, Institutions, Companies and Trusts' were able to join the CMFA. Individual membership was $3 a year; other organisations $20. The National Council was to consist of the president, life members, six vice-presidents preferably one from each State, and 12 councillors preferably with representation from each State, along with 'honorary', i.e. unpaid, secretary and treasurer.

This constitution would eventually cause headaches as naturally NSW based officers came to dominate the National Council. To the States as they organised, NSW *was* the National Council and vice versa. Naturally with the life members on Council who had a vote as well, the tone was set from the beginning of NSW dominance—and members were exclusively Army. At the end of 1972 Cullen made a very brief report to the members:

> The year just concluded has again achieved progress. Our financial situation remains satisfactory, within the limits of our requirements. In my opinion, we are fulfilling our function, not only by our conscientious efforts, but by our existence.[13]

Actually, the balance sheet in June 1972 was $610.40 in the red and still $120 in the red the following year within the context of less than $3000 in turnover, mostly expended on production of *Citizen Soldier*. A sounder footing was established in 1974 with a fund raiser—the 'Guard's Concert' which raised nearly $5000 and then, in 1977, with over $7000 in donations. Subscriptions by members—at $5 a head—was $1440 in 1972, implying nearly 300 members. By 1977 subscriptions had dropped to $290.[14] Here we see the mixing of the NSW branch subscriptions with the National Council finances as the subscriptions related entirely to the branch membership.

The reality behind the modest report and growth to follow through the 70s was the dominant leadership of one personality, that of Paul Cullen. He was a 'staunch and combative advocate of CMF interests both within the army and, more particularly, outside of it, including in his ability and willingness to lobby directly at the political level'.[15] In the first two years of the Association, the constitution needed only five for a quorum (and thereafter nine). That indicated that the CMFA was hardly a well-organised entity straight from the womb and probably recognised that to get anything done, Cullen only needed his closest associates on the Council in support. However, by August 1971 the CMFA boasted 700 members, or so it said.[16] In his introduction to the first edition of *Citizen Soldier*, Cullen wrote:

> Today the CMF is again at the crossroads...Today, with the withdrawal from Vietnam and the reduction of National Service from two years to eighteen months, and the ALP platform to eliminate National Service...the whole situation is in the melting pot. It is within the circumstances created by this confusing background that our Association looks at the situation of the [CMF].[17]

The second National Service scheme was introduced by the Government from 1964 to enable the Army's Royal Australian Regiment (RAR) to expand from four to eight battalions to meet its engagement with Indonesia's 'confrontation' campaign and subsequent operations in Borneo and Malaya. The Vietnam war then involved Australian forces from 1965 with an ever-growing commitment, and so the scheme was continued. The scheme had the effect of the CMF 'losing many of its more experienced and motivated soldiers' to active service but receiving in turn 'disgruntled [National Service] recruits' who had the option of the CMF if they did not want to volunteer for active service. In turn it was abolished by the Labor Government in 1973.[18]

Cullen went on to note in *Citizen Soldier* that 'public attraction to voluntary service in the CMF is declining as proven by the ever-reducing strength of the CMF...[which] proves this point'. Cullen hoped that an extension of the National Service ballot could be extended to the

CMF on the basis of 'five years of service and a 28 day camp', with the aim of achieving 50,000 members.[19] He knew that without these sorts of numbers, the CMF could not be efficient. Although the CMF had 'been removed from the possibility of operational service', during the National Service period of the Vietnam War, with supplementary conscripts the CMF had probably reached the highest level of efficiency (i.e. through high numbers) it had been in since 1948 despite very high turnover of all ranks during the period.[20]

Somewhat optimistically, the first editor of *Citizen Soldier*, Lieutenant Colonel Tony Browne—and a later president of the Imperial Services Club—wrote: 'With the current withdrawal of our Regular Forces from Vietnam and the forthcoming reorganisation of the Army, now is the time for rethinking the ways in which the CMF can again become *a force to be reckoned with*'. [Browne's emphasis] [21] However, the efficient near full strength CMF of the early-mid 1960s did not last, and indeed National Service had in effect made the CMF redundant within the context of its former role as 'the 3rd AIF in being', in case of a major conflict.

Now, in the heightened uncertainty following the Vietnam war, what was the CMF's role? Cullen had tried to have a CMF battalion accepted for service in Vietnam, both when he was CMF member on the Military Board (1965-66) and afterwards as president of the newly formed CMFA, but he had been unsuccessful. He even advocated raising the CMF strength to five divisions complete with a corps headquarters by 1975, but this was completely unrealistic in the financial and political circumstances of the day. So 'the CMF entered the 1970s under a cloud of uncertainty'.[22]

This sense of drift however was the very thing that Cullen would work steadfastly to resolve: '...he was probably happier outside the chain of command (although he had rarely, if ever, allowed it to restrict his freedom of action as he saw it)'.[23] He wrote:

> ...we can see our task, entrusted to us by our members from all States of Australia, is to put forward our ideas in order to make the CMF the efficient force of adequate size, properly equipped, that we deem to be essential, in support of the Regular Army for the short and long term defence of Australia.[24]

At the beginning the Association was not taken very seriously and was viewed with a good deal of suspicion by ARA and serving CMF members alike. In early July 1970 Cullen had written to the Minister of the Army claiming that branches had been formed in NSW, Victoria, and Queensland. Cullen's letter had been mentioned at the Military Board by the Secretary, Department of Army. The then CMF Member of the Military Board, Major General Norman A. Vickery CBE, MC, ED (1966-70—he had replaced Cullen in that role), felt compelled to respond to Cullen's assertions.

In a letter dated 26 August 1970, Vickery noted that he had become 'increasingly aware of growing concern among serving officers (particularly CMF officers) throughout Australia, at the fact and mode of formation of the [CMFA], many of its expressed aims and its activities...' Vickery pushed back against Cullen's claims of other branches being formed: 'Enquiries revealed that only one branch (NSW) in fact existed...an approach made to a senior retired officer in a Command, even if he expressed general agreement, does not amount to the formation of a branch.'[25]

Vickery informed Cullen that after consultation with the Chief of the General Staff (CGS) and the Advocate General, he had raised the whole question of the Association with the CMF officers assembled for the CGS Exercise on 9 August and discuss it at the CMF Conference on 14 August. Major General Murchison, as the CMFA NSW president, participated in that discussion.[26] Vickery summarized the findings, including:

> It was unanimously stated that there was keen concern at the dangers inherent in some of the aims of the Association and as evidenced by its recent activities, in applying external pressure in major or minor matters affecting the Army CMF [ACMF]...these aspects they saw as potentially divisive and damaging.

> No serving officer should pass to the Association information other than that relating to the general field of recruiting, PR [public relations], etc. E Comd [Eastern Command] officers felt that serving officers should not sit on any Council of the Association. All others felt that serving officers should not join it.

> ...Some Commands acknowledged that there was the grain of a good idea behind the formation of the [CMFA] for work in the fields of recruiting, PR and the improvement of the CMF image in particular...However, others expressed the view that such activities could be promoted within existing organisations, while others felt that State organisations on a broad base should be established for this purpose, but not on a national basis as yet.

> W Comd [Western Command] expressed the forthright view that it was an impertinence for a National Council to have been formed in the manner it had been and invite the formation of State branches as envisaged in the constitution...and considered wrong for the National Council to put forward publicly as an ostensibly representative body.[27]

Vickery added that 'As these views flow from the assembled senior representatives of the ACMF they must be given great weight'. He warned Cullen of the consequences of his position as CMF Member and its relationship with Army HQ being undermined by the activities of the Association especially when Cullen was referred to in the press as 'the head of the CMF'.[28] Vickery's letter was copied to the Minister, the CGS and the AG. Major General Gordon L. Maitland, OBE RFD ED and at the time commander of 2nd Division, later commented that when Cullen established the CMFA '... initially he didn't have anyone in it! So for a while he *was* the CMF Association'.[29]

Nonetheless Paul Cullen—who had retired in early 1966 after almost 40 years of Army service including in North Africa, Greece, Crete and New Guinea during World War II—was not to be deterred. His forceful and determined personality was harnessed to the full in his advocacy as president of the CMFA, of the CMF and its place in defence planning while at the same time continuing every effort to secure better conditions of service for individual CMF members. He would continue to play a major role in the Association and on behalf of reservists for more than 20 years.

The CMFA National Council (such as it was), according to later notes by Major General Warren E. Glenny, AO RFD ED, had already made its first submission though Cullen. In 1970, Major General Francis ('Frank') G. Hassett, CB, DSO, LVO was selected to lead the Army Review Committee, the so-called 'Hassett Committee'. The committee's far reaching reforms included moving from a geographical to a functional command system. This involved (in part) the replacement of the various State Army Command Headquarters with a national field force, training and logistics command system.'[30]

It was among several reviews and changes affecting the CMF before the major Millar Report of 1974. Hassett would have been surprised to receive a submission from the new CMFA and according to Dayton McCarthy, could not even recall the CMFA's existence when he was the CGS.[31] Major General W. Glenny later said the Association even adopted as policy the concept of a Ready Reserve for the CMF at the meeting in February 1971.[32] But in these early days of the Association, it is not clear whether such resolutions by the small group of CMFA officers at the National Council in in NSW even came to the real attention of the Ministers and senior serving officers let alone influenced their thinking in any way.

By 1970 the CMF had not recovered organisationally from the disastrous Pentropic Division 'reforms' of a decade before.[33] These 'reforms' were later abandoned in 'some tacit recognition that the changes inflicted on the CMF had done serious, lasting structural damage'.[34] CMF numbers had then been boosted from 1964 by the entry of National Servicemen who preferred to serve six years in the CMF rather than be sent to Vietnam during a two year ARA duty. In turn, this allowed the CMF to train with full complements of staff and soldiers in most cases and therefore be deemed 'efficient'. However, it had not been able to serve in Vietnam with formed CMF units and the National Service scheme supplemented the ARA rather than the CMF (although numbers of CMF did join the ARA during the period). When the war

Melbourne University Regiment reservists on exercise c1971
(MUR Archives)

ended and the Whitlam (Labor) government halted National Service, CMF numbers halved virtually overnight, crashing from 51,000 to barely 20,000.[35]

The strategic landscape had been changing for some time, and the CMF was found wanting, with many officers still caught in the paradigm which had determined the CMF's national expansion role—that it would fill out the Army in a general mobilisation in event of a general conflict. After the Vietnam war, it was inevitable that a reorganisation would be coming as the Army itself restructured and that expansion paradigm shattered, perhaps forever. 'It must be concluded', wrote one historian, 'that the years the CMF spent on the home front during the Vietnam War were the most destructive in its existence to its public standing and to its own morale.'[36]

Some would say that it has been in steady decline since then, but 'hindsight is a wonderful thing' and the future for the CMF/Reserves held many positives. However, the CMFA and its successors, the Army Reserve Association (AResA), and the Defence Reserves Association (DRA), would find itself in many battles in the years ahead protecting and preserving the great legacy of the *Citizen Soldier*. The Association also needed to adapt to the reality of the political, financial, and strategic defence evolutions in the years ahead to help maintain the relevance and sustainability of the Reserves.

With election of the new Labor government and the end of National Service in December 1972, also came an announcement of an inquiry into the CMF headed by Dr Thomas B. ('TB') Millar. The focus in 1973 of the Association's representations was to that committee of inquiry (the Millar Committee) as well as to the Senate's Standing Committee for Foreign Affairs and Defence, and the Defence Committee of the Liberal Party.[37] The then Minister for Defence,

Lance H. Barnard (Labor/Bass)—who had seen service with the AIF—wrote in the October 1973 edition of *Citizen Soldier*:

> The [Senate Standing] Committee will look in particular at the military capabilities and specialist support which the CMF can provide to the Australian Regular Army....Most importantly, the Committee will make recommendations about the definition of the role of the CMF. Until that is clear we cannot decide on its status or responsibility.[38]

Although some CMFA members continued to hope that another National Service scheme would swell its numbers once again, the Association had to focus on what it could influence. That year, Cullen wrote down what he believed were the basic and vital issues that the Association should be focused on, noting that these points were the core of a longer submission from the CMFA to the Millar Committee:

1. It must really, really be One Army—really and truly—with the same role and task at different stages
2. Change the name from CMF to Army Active Reserves
3. Clothing to be identical. No little savings please. (Don't spoil ship for h'aporth of tar etc.)
4. Pay rates also (Note: Regulars 5 days, AAR 7 days)
5. Barracks of equal standard
6. Messes and canteens of equal standard—up to Club standard and staffed.
7. Overseas training every three years for efficient soldiers.
8. Efficiency Grant every fourth year = a colour TV set or equivalent. Bring Mum in to support.
9. Efficiency Grant for female soldiers.
10. More autonomy and discretionary authority for Unit and Sub-Unit Commanders in recruiting and some training aspects.
11. Closer liaison with regular units with close links and shadow postings with them where appropriate.
12. Ready Reaction Force to be made ready with adequate conditions and inducements.
13. ARA cadre to AAR to be the cream, like it used to be, when for example, Lt-Gen Sir Thomas Daly was the [Adjutant] of a Light Horse unit and wanted to be.
14. An Army Technical College with correspondence courses in each command for specialists, NCO's promotion etc.
15. Development of the Bushmen's Rifles system of mainly camp only activities.[39]

Glenny would later wryly note that 'Readers with an interest in Defence issues will feel that they have entered a time warp as they read of those issues that attracted the attention of the Association in its early days and as they read ... of its evolution into the DRA will see that whilst much changes, much remains the same.'[40] However at the time, with Cullen as a forthright and well-connected leader, it seemed perhaps to Cullen that all of those things on his list were achievable, especially if there was 'real and believable support from our leaders—Governor-General, Prime Minister, Defence Minister, and everyone'. He added 'Here's hoping!'.[41] Cullen's hopes extended to many in the CMF, who saw the inquiry as something that might get the CMF back to its former status.

The *Citizen Soldier* magazine October 1973
(DRA Archives)

The Millar Committee Report was completed in March 1974. In the long list of submissions to the inquiry—over 1000—only two listed were formally from the CMFA (from the CMFA—NSW and WA branches), and others from Paul Cullen, presumably incorporating the points made by Cullen in *Citizen Soldier*, along with over 100 from officers with an

Efficiency Decoration (ED) after their name and hundreds of others who can be assumed to be current or former reservists of all ranks. Identifiable among them were numerous officers later to be involved in active roles in the CMFA, especially from Victoria. In its conclusion the Report said that 'some of the recommendations...will be controversial...among the [CMF] and even with the public at large...But the world has changed in the nearly 30 years since the end of World War II, and the needs of defence have changed'.[42]

Major General Cullen later wrote a long article in the then *Pacific Defence Reporter* which praised the report as thorough, professional, and factual, constructive and forward thinking—he noted that he disagreed with 'a few, very vital basic concepts. But I do emphasise that they are the vital points.'[43] Years later, CMFA branches were still engaged in discussion, sometimes heated discussion, over the Millar Report and its consequences. Some older members were still trying to turn back the clock; others were trying to understand how the 29 principal recommendations of the report impacted on the CMF—now to be called the Australian Army Reserve.

> Far from being a vehicle of the Regular Army to denigrate the CMF as some opponents predicted, the report did much to highlight many of the conceptual and structural problems that the CMF was afflicted by at the time, however, the way in which the government chose to implement the recommendations, and indeed the way in which some of them were allowed to lapse, ultimately served to at least partially justify some of the cynicism voiced in certain CMF circles about the report.[44]

Nonetheless there was a wide range of reactions to the Millar Report. As is usual with Government take up of such inquiries and reports, some of the recommendations were accepted immediately; others were not. Many in the CMFA saw the Government taking up the least popular recommendations at the expense of the CMF; others saw Regular Army machinations at work. Some recommendations took years to implement, and others were never implemented. One example was the name change to Army Reserve—the CMFA itself did not change its title to reflect this until late 1986. The Victoria-based 3 Division was especially affected—it was downgraded through amalgamation to become 3 Division Field Force Group with a brigadier in command.[45]

> Like the Pentropic reorganisation, the implementation of the Millar recommendations caused tremendous pain across the CMF. Certainly some pain was unavoidable... [however] the ARA showed little sensitivity or understanding of the effect that wholesale elimination of units would have...[and] the army did not then move forward with any positive steps that would revitalise the morale...and lay the basis for a viable Army Reserve. [The recommendation for] a total force...did not occur, and the ARA saw the reform more of a way to subsume the reserves instead of helping them to grow.[46]

Cullen saw the recommendations as good and bad—'in retrospect, many former CMF members believe that far more of the bad recommendations were implemented than the good ones.'[47] Unit amalgamations were seen as one of the 'bad' recommendations—and by the end of June 1975, 17 major and 246 minor Reserve units had been involved in reorganisations of some kind.[48] By 1976, the Military Board and CMF member position had gone; in its place the CGS with a Chief of Reserves (CRES)—Major General Bruce A. McDonald DSO OBE MC, a Regular Army officer—as advisor on Reserve matters. By then, a new Government was in place—Liberal, with different priorities.

In early December 1974 Prime Minister E. Gough Whitlam MHR (Labor/Werriwa), attended a CMFA luncheon and listed the Government's implementation of the Millar Report recommendations. The Leader of the Opposition, J. Malcom Fraser MHR (Liberal/Wannon), made a strong reply, stating in *Citizen Soldier* that year that there were 'great and fundamental differences' between the Government and Opposition with regard to defence planning, accusing the Government of being 'irresponsible to have such naive confidence in the future that we can say there is no danger to Australia for 10-15 years.'[49]

The RSL chimed in as well. Colin J. Hines (later Sir Colin Hines OBE) NSW State President of the RSL and member of the CMFA NSW Committee found empathetic CMFA spirits when he declared, 12 months after the Millar Report was tabled, 'the new Reserve still languishes with all the old problems of the CMF. It lacks soldiers, instructors, regular staff, modern equipment, administrative support and Government direction.'[50] It needs to be remembered, to put these comments into context, that most of the senior CMF and ARA officers at the time had not only seen first-hand what a real threat to Australia looked like in World War II but also were now living through the Cold War with all that that entailed.

Within the Association, changes occurred when Cullen stepped back from president at the National Council meeting of 10 March 1975 although remained on the Council as immediate past president. His successor, Major General Murchison, said:

> The Association owes a great deal to General Cullen who, as foundation President, has kept the momentum going and the organisation alive and very much kicking—despite the lack of support from many quarters from which one might have expected it. Murchison noted that Cullen's expertise, drive and energy will still be available despite the great demands it makes of his time.[51]

The CMFA National Council in place when the Fraser Government came to power in December 1975, with some other changes to the original coterie around Cullen, consisted of:

Life Members
>   Lieutenant General Sir Edmund Herring
>   Major General Sir Victor Windeyer
>   Major General Sir Denzil Macarthur-Onslow
>   Major General Sir Ivan Dougherty
>   National Council
>   President: Major General Murchison
>   Past President: Major General Cullen

Councillors:
>   Lieutenant General Sir Mervyn F. Brogan
>   Brigadier Sir John Pagan
>   Sir Colin Hines (also president of NSW RSL)
>   Colonel Ronald S. Garland MC*
>   Treasurer: Lieutenant Colonel Ronald M. Faulks ED
>   Secretary: Colonel Dart
>   Editor *Citizen Soldier*: Lieutenant Colonel A. Browne
>   Publisher *Citizen Soldier*: Major B. Nebenzahl

State Presidents:
>   New South Wales—Colonel T. J. Crawford ISO MBE ED[52]
>   Victoria—Brigadier Ian H. Lowen OBE ED
>   South Australia—Brigadier Sir Thomas Eastick
>   Western Australia—Brigadier J. B. Roberts MBE ED
>   Tasmania—Colonel Lionel A. Simpson CBE
>   Queensland—Brigadier Malcolm E. Just OBE ED
>   Northern Territory—Captain Jeff Dunn[53]

As will be seen, Cullen would continue to be very much involved in the CMFA for years to come. Murchison had plans as well. Not least was that he wanted to hold the first national conference of the CMFA in Melbourne on 3 August 1976.[54] He commented favourably on D. James ('Jim') Killen MHR, as the Liberal Shadow Minister for Defence, believing that he would prove to be 'an able and energetic Shadow Minister'.[55]

One reason for the comment was that, as Cullen had pointed out in a recent letter to Association members, 'simple things—such as the issue of boots, polyesters etc. have for no apparent reason simply not been implemented'. Murchison felt that 'it should be possible to do something, if only by turning on the spotlight to help cut through whatever barrier exists to the implementation of this and many of the other recommendations of the Millar Committee'.⁵⁶ All hoped Jim Killen, a former RAAF NCO with World War II service, would do better for the CMF than Lance Barnard.

2.
Wax and Wane

By the end of 1976, the Association was gaining strength. At the National Council meeting in Melbourne in August 1976 several well regarded Reserve officers in Victoria were invited to attend along with the Victorian State president, Brigadier Eason. These included Brigadiers W. H. 'Mac' Grant, Keith V. Rossi and Francis E. Poke, RFD, ED along with Lieutenant Colonel David E. F. Bullard—more attended the reception afterwards.[57]

It was also reported by Captain J. Dunn in the Northern Territory (NT) that a branch meeting would be held in January 1977 when news might be available concerning the re-establishment of the Army Reserve Independent Company in Darwin. The NT branch of the CMFA at the time reported that it had 41 members but there was no reserve unit in the NT. Colonel Lionel Simpson also relayed to the meeting that a Tasmanian branch would also be formed early in 1977.[58]

Matters discussed at the National Council meeting that year included long service awards, a function for Prime Minister Fraser, in Canberra, and the current reorganisation still underway arising from the Millar report. The Council also discussed at length travelling allowances for Army Reserve members and the Council resolved to make strong recommendations to the Minister for Defence, Jim Killen, to request that the allowance be re-introduced and made retrospective to 1 January 1976. The Association also wanted to be represented on the new Committee of Employer Support Army Reserve (CESAR).[59]

In November 1976 a reception was held at the 3 RNSWR Training Depot in Canberra, attended by the Prime Minister, Minister for Defence (regarded as especially sympathetic to the Army Reserve) and senior defence force officers such as the new Chief of Defence Force Staff (CDFS) General Sir Frank Hassett and the CGS, Lieutenant General Arthur L. MacDonald CB. Major General Murchison, CMFA president, and Major General Cullen attended.

Most important for the CMFA, at the reception the Prime Minister made a lengthy statement of support for the Army Reserve, which came just two weeks after the Government's White Paper had been released.

Major General Cullen holds court with Minister of Defence Killen, (L), Prime Minister Fraser and Major General Murchison at a CMFA reception 1976.
(*Citizen Soldier*, December 1976)

Malcolm Fraser said:

> Reserve Forces are a fundamental part of the force structure and they have an important place in our basic approach to defence...we place real weight on improvement of the effectiveness of the Army reserve...we recognise the vital contribution which the reserve makes to our defence capability, and we will work to ensure that it maintains its capacity and strength.

He also noted that the Government did not support defence service homes for reservists, and, while attracted to the idea of legislation as recommended in the Millar report to allow the Reserve to be mobilised for limited periods of operational service outside of Australia, that this was 'not a step to be rushed' and that 'it would be a pointless exercise if it was not accepted by the community, by employers and by existing and potential members of the Army Reserve'. Fraser did, however, announce an increase in strength by 5000 over the following five years—an increase on current strength of 25%.[60]

Subsequent years saw the Association making representations on a wide range of issues a number of which flowed from the Millar Committee recommendations, positive forward-thinking recommendations, and the Association's concerns as to various decisions seen as challenging to the future of Defence Reserves. 1975 and beyond saw the Association heavily involved in representations to Government and Defence as the implementation of Miller recommendations took effect. Issues addressed included:

- The need for a stated role for the Reserve
- Action to establish a Committee for Employer Support of Reserve Forces
- Need to delay closures of Army Reserve Depots
- Concern as to delays in implementing of the positive aspects of pay, housing loans and conditions of service
- Problems with Long Service Awards and the lack of Honours and Awards for reservists.
- The issuing of Identity Cards to Reserve members
- If the Efficiency Decoration and Efficiency Medal could not be re-introduced, then a Reserve Forces Decoration and a Reserve Forces Medal should be introduced to replace them.
- The need for legislation enabling Call Out of the Army Reserves.[61]

Following the 1976 National Council meeting it was agreed to form a Victorian branch of the CMFA—its first formal meeting was held on 8 March 1977. Their first committee met at the venerable Naval & Military Club in Melbourne. Chairing the meeting was Brigadier Ian Lowen who later would also become the CMFA president. Lowen had already travelled to Sydney to meet members of the National Council and after to Canberra in November 1976 for the reception there with Prime Minister Fraser. Among those attending the first Victorian meeting were Lieutenant Colonel Don H. McLeod ED who would later become CMFA secretary; Lieutenant Colonel David Bullard who would later succeed Lowen as president of the Victorian branch; and by invitation, Major General Francis Poke, at that time the senior CMF officer in Victoria.[62]

The formation of the Victorian branch had been a long time coming and the pathway taken to its formation is a good example of how tentative the idea of a CMF association could be. Following the now Brigadier Dart's letter to Major General S. M. McDonald in 1970, inviting him to form a Victorian Committee. McDonald wrote in turn to Major General Kenneth D. Green OBE ED, then Commander 3 Division, asking him to suggest a meeting along with the proposal that Lieutenant Colonel Norman Wright ED could be secretary.[63] McDonald convened the meeting at the officers' mess of 3 Division (at 'Grosvenor' on Queen's Road in the city, where many subsequent CMFA meetings would be held) on 2 July 1970 with the purpose of 'obtaining the advice and encouragement of those present in establishing an organisation in Victoria to foster and assist in all possible ways the [CMF] as an integral and significant part of the Australian Army'—44 attended the meeting and 21 sent apologies.[64]

Then, perhaps because they remained unsure of the standing of this new idea for a CMF association, the meeting made the decision to ask the General Officer Commanding (GOC) Southern Command, Major General Colin A. E. Fraser CBE (the most senior Regular officer in the State) to convene a meeting to establish a Steering Committee charged with reporting back to the initial group. That meeting was held 28 August 1970 and immediately started to move away from Major General Cullen's original objectives. This was perhaps unsurprising due to the deference to the GOC. A year and six meetings later the 'Steering Committee' reported back to announce that the 'organisation and aims proposed by Major General Cullen *was not quite what was wanted in this Command*' [author's italics].[65]

The Steering Committee proposed a policy committee led by GOC Southern Command, with Commander 3 Division, the State president of the RSL and a senior retired officer (Major General McDonald). Major General Vickery, recently retired as CMF Member on the Military Board, pointed out that what had become a Southern Command organisation was designed to support the Army as a whole and not the CMF in particular as proposed by Cullen and the NSW organisation. Major General Fraser conceded that the committee should not be led by a Regular Officer and Major General McDonald was elected as the head of that committee.

Unfortunately, the records do not survive to show what happened between the meeting on 25 August 1971 and the formal formation of the CMFA branch in Victoria in March 1977, but clearly matters fell into abeyance despite the apparent enormous interest at the beginning. A later set of CMFA Victoria committee minutes simply noted that 'the forerunner of the CMFA in Victoria was the Defence Forces Advancement Group'. More pointedly, in the Victorian Committee's first annual report the new Victorian branch president, Brigadier Lowen said:

> ...I would like to acknowledge that I am aware that the CMFA, for reasons that will not be explored here, did not get off to an auspicious start here in Victoria. I believe this was unfortunate for both Victoria and the national body whose voice was weakened, in that until now it was not completely national.[66]

This glimpse at the whole story illustrates the challenges which faced Major General Cullen's ambitions for the CMF Association in Victoria and no doubt elsewhere as well.[67]

Turning briefly to RAAF Reserves, during the 1970s the policy of affiliation and co-location of Citizen Air Force (CAF) and Permanent Air Force (PAF) squadrons had on the whole been successful. The increasing level of joint operations with Navy and Army, moreover, meant that there were more Air Force Reserve specialists required, such as Operations Officers, as well as the various professional staff such as doctors, lawyers, and dentists. At the same time, in 1973, university squadrons were disbanded, and recruiting also ended for the Air Force Emergency Reserve, decisions that were 'probably influenced 'by the anti-defence sentiment that was building up in the ...1970s'.[68] For the CMFA, however, focusing on its own branch development and Army Reserve matters in the 1970s, Air Force—and Navy—Reservist concerns received the least attention from the Association.

In 1976, the Government announcement that the replacement of Imperial Awards with new Australian Order medals and awards had neglected to include new Reserve service medals in the Australian system. In this and meetings to come in 1977 throughout the CMFA an immediate major issue was to persuade the Government to replace the Efficiency Decoration with a new medal under the new Australian Order. An interesting discussion also took place at that first meeting in Victoria—'senior ex-Australian Regular Army (ARA) officers are being transferred into the CMF and were dislocating the career plans of CMF officers and stifling opportunities for promotion'. It was perhaps a typical concern reflecting the mistrust the CMF officers had in the ARA and this would be a theme repeated for many years ahead.[69] Soon after, the CMFA held its AGM in Sydney on 18 March 1977. Records of that meeting have yet to be found.

A major development of great interest to the CMFA in 1977 was the establishment of the Committee for Employer Support of Reserve Forces (CESAR), arising from the Millar Report. CMFA State presidents were encouraged to nominate as CMFA representatives on CESAR committees; the CMFA president, Major General Murchison, joined the national CESAR committee. CESAR intended to deal with the top 100 companies in Australia to enlist their help to support their employees who were reservists to attend camps and training without loss of employee benefits.

Typical of the initial activities were CMFA sponsored receptions for Federal politicians, State premiers, Commonwealth and state public servants, employers and employer bodies and the Defence Forces both Reserve and Army. Later that year, CESAR was renamed the Committee for Employer Support of Reserve Forces (CESRF).[70] Involvement with CESRF became an important part of the CMFA's action agenda over coming years—Major General Cullen himself would later head the CESRF committee in NSW.[71]

In early 1978, Murchison went to Canberra to discuss among other issues on the CMFA agenda, Reserve awards. A submission supporting retention of existing awards was sent to the PM and all cabinet ministers. In the meantime petitions were being sent to members

1/77 UNSWR Subj.2 Cpl Course, May 1977
(*The Kensingtonian*, February 2016)

of parliament—by April 1978, 61 petitions had been sent to the House of Representatives and 38 to the Senate.[72] Typical of these was the following, submitted to the House of Representatives on 24 October 1978:

Citizen Forces: Long Service and Good Conduct Medals

The Honourable the Speaker and Members of the House of Representatives in Parliament assembled. The humble petition of the undersigned members and ex-members of the Citizens Forces of Australia respectfully showeth:

(1) On 14 February 1975, the then Australian Government deprived the Officers and men of the Australian Citizen Naval Military and Air Forces of the distinctive and historic decorations and medals for long service and good conduct, namely the Reserve Decoration, the Efficiency Decoration, the Air Efficiency Award, the Efficiency Medal and Long Service and Good Conduct Medals, awarded for long and meritorious voluntary service in the citizen forces;

(2) The proposed substitution of the National Medal for these decorations and medals varies the principle of selective recognition of efficient voluntary service in the citizen forces in that it recognises the period of service only and embraces also full time service as well in the defence forces as in the police, fire brigade and ambulance services;

(3) This deprivation caused and is continuing to cause serious discontent amongst personnel of the citizens Forces who willingfully and cheerfully give of their spare time

outside their normal full time civilian careers, to serve Her Majesty and Australia;

(4) The Reserve Forces of Australia have been recognised by the present Government as a valuable and cost-effective component of the Defence Forces. Anomalously, whilst the Government is actually supporting recruiting for these Forces it has imposed and continued this deprivation which as foresaid has depressed the morale of the Citizen Forces;

(5) Her Majesty has not cancelled the said Decorations and Medals.

Your petitioners therefore humbly pray your Honourable House take appropriate action to resume the award of the several distinctive Reserve Forces Decorations and Medals for Long Service and Good Conduct to members of the Royal Australian Naval Reserve, Army Reserve (CMF) and the RAAF Citizens Air Force.[73]

The covering letter to the submission, in an early example of tri-service cooperation, was signed by Lieutenant General Sir Edmund Herring, Lieutenant Commander Sir Neville Pixley CMG MBE VD KtON, RANR and Air Chief Marsal Sir Frederick R. Scherger KBE CB DSO AFC.[74] This was a prime example of how the CMFA was able to bring to bear exemplary 'influencers' to support the common cause. The biggest flaw in the CMFA argument, however, was that the CMFA was fighting to retain the old awards rather than reading the Government's inclination to accept the previous Labor Government's intention to create new Australian awards.[75]

Despite considerable sympathy to the CMFA's standpoint within Government, the mood of the times was against the CMFA in that regard. Several years later, when awards were finalised, the Reserves were indeed recognised with the new Reserve Forces Decoration (RFD) and Medal, and this was a testament to the consistent lobbying and efforts by the CMFA to have Government recognise Reserve service. In a similar, in hindsight, somewhat quixotic fashion, the CMFA resisted the new, but not formally adopted, definition of the role of the Reserves arising from the Millar Report in 1974. The definition emphasised the complementary nature of the regular and reserve components of the Army:

- The Army reserve, as part of the Australian Army, is to participate in the defence of Australia and its interests in times of war and defence emergency.
- The Reserve, with the Regular Army, is to provide the basis for expansion of the Australian Army.

The CMFA certainly did not want the Reserve to be simply seen as an 'appendage' of the ARA.[76] Yet while the sentiment is understood, was this something the CMFA really thought it could change? Nonetheless, Major General Cullen tried to get some definition on this from Defence, for the aims of the CMFA itself rested on a clearly defined role for the Reserve.

As the Association reviewed these aspects it remained unmoved by the official change of the name CMF to Army Reserve—it would be several more years before the Association was to change its name to reflect this. A similar change occurred in RAAF where following a RAAF-wide review in 1978 the generic term RAAF Reserve was adopted to cover all Active, Specialist and General elements.[77]

The CMFA also engaged closely with the 'Coldham Committee'—the Committee of Reference for Defence Force Pay, under the judicial chairmanship of Justice Peter A. Coldham—to have Reserve pay carefully considered as well as regular.[78] It was also concerned with Reserve retention and the recruiting process—a psychometric test had been introduced before attestation that had lengthened the time for the recruiting process to be completed, with recruits dropping out because of the time to process them. Trade testing and pay rates for specialists and the effects of financial structures and administration red tape on the Reserves also came under the attention of the CMFA. By mid-1978 Reserve numbers had fallen to a new low of only 22,000, compared to the high point of 35,590 in November 1968.[79]

The National Council was also looking at the question of a Reserve Staff & Command College being established. A report on the question by Brigadier Owen Magee, a former regular officer who had joined the Reserve in 1970, was to have been released in January 1978 but several months later it was still with the National Council.[80] It was later submitted to Major General B McDonald of Training Command. By the time of the August 1978 AGM of the National Council in Sydney, many of these matters were unresolved; the aims of CMFA were not discussed and role of CMF was deferred to the next meeting of the National Council.[81] One important change was the appointment of Colonel Glenny as Secretary vice Brigadier Dart who had been promoted and appointed Commander of 2 Training Group. Glenny would later become a major general and Association national president.

The AGM was addressed by Major General Maitland, at the time Inspector-General of Reserves—and later a national president of the Association.[82] He noted that the 'Millar Report was now history and its recommendations a dead letter', although it is not clear now why he asserted this at the time. Maitland said that the destruction of units would cease but he also noted that several units were on the line with low numbers. The proposed Call Out Legislation was 'bogged down' in in Defence Central. He also said that 'neither he nor anyone else knew the current role of the CMF'.[83] Meanwhile the campaign for the resumption of 'traditional awards' continued. The CMFA enlisted the support from the new Australian Defence Association, the RSL and the Veterans and Defence Service Council to the commanding officer of 2 Battalion, Royal Victorian Regiment (RVR), invoking the Scottish links of the unit to PM Malcolm Fraser.[84]

The National Council meeting also agreed to press for the elimination of psychometric testing from the recruitment process. It also decided to approach the appropriate Government departments to discuss suspension of unemployment benefits during camps of continuous training and the need to serve another re-qualifying period afterwards. And it would take up the whole subject of CMF administration with Defence Minister Killen.[85] Meanwhile, in an indication of the sometimes-hesitant nature of the national organisation of the Association and its funding, *Citizen Soldier* once again went into abeyance, to be replaced by newsletters.[86]

The Victorian Committee was active in following up the National Council on matters discussed or agreed at the last CMFA AGM. What is clear is that the Victorian chairman, Brigadier Lowen, had no hesitation to writing directly to anyone whom he thought could help the Association. This included the PM, Ministers, individual politicians, serving senior officers both Regular and Reserves including Service Chiefs, and of course the National Council itself. Whether this was because he saw the National Council as lethargic or simply wanted to assert the Victorians against the NSW-centric CMFA leadership is not known.

It did, however, cause some friction with the National Council as will be seen and some frustration around communications between the two bodies started to become evident. For example, it was nearly 12 months following the August 1978 AGM before the National Council was able to elicit a definitive response from Government re unemployment benefits and camp attendance. By then, in mid-1979, a new front had emerged—to change the *Compensation (Commonwealth Employees) Act* 1971 to improve pay outs for Reserve members injured on duty.

In addition to the AGM held in Sydney on 14 September 1979, the National Council had also met in Brisbane in March, according to its annual report of late 1979. The report also noted that the Executive met regularly between Council meetings to conduct the daily affairs of the Association.[87] While the annual general meetings were the summary and progress meetings of the Association, it is useful to be reminded of the constant exchange of correspondence, reports, papers, views and submissions of All kinds which flowed in and out of the executive of the National Council, remembering also that this was in the age before computers, email, and mobile phones. That year's AGM saw a change in presidency from Major General Murchison to Major General John Mc. L. Macdonald, AO MBE RFD ED.[88]

Among the matters discussed at the National Council meeting after the AGM, at least unemployment benefits as an issue was laid to rest with the administrative restrictions eased for Reserve members to enable advance claims before camps. Perhaps buoyed by that success the National Council resolved to 'take up with vigour' the matter of adequate compensation for the Reserve, especially employer liability insurance. The issue of Call Out Legislation, however, remained far from resolution—and it would prove to be one of the longest running issues ever faced by the Association, given the multi-faceted aspects to this somewhat complicated question. The Association remained confident that its campaign for the re-introduction of traditional awards was won. The CESRF again factored large in the deliberations. It was not only supporting a reception with the Governor-General Sir Zelman Cowen and Lady Cowen in November 1979 but also about to launch television advertising in support of Reserve Forces.[89]

In a somewhat embarrassing revelation, the National Treasurer had reported to the meeting that of $3774 in the National Council's funds, $3309 had been transferred to the NSW branch 'for the purpose of re-establishing a viable State branch and to promote a memorial to the [RNSWR]'.[90] While the Association belatedly bowed out of the memorial proposal, nonetheless the transfer of assets was deemed to be in error and was corrected by the Treasurer. The representatives of the State branches at the meeting would not have looked upon this transgression with any amusement given it was as much their capitation money in the funds as the National Council's. No doubt eyebrows would have been raised at these 'goings-on' in NSW.

The 1980s saw strong representations by the Association regarding:

- The re-raising of the 2nd and 3rd Divisions from the existing Field Force Groups.
- The re-raising of Brigade Headquarters.
- The introduction of Reserve Command and Staff Colleges.
- Overseas training for reservists.
- Measures to improve retention.
- The failure still to introduce the recommended legislation for Call Out of Reserve Forces and importantly Protection of reservists on Call Out.[91]

Numbers in the CMF were rising again with an authorised increase of 8,000 and a target of 30,000 was set for mid-1980.[92] Meanwhile, the Victorian branch led by Brigadier Lowen was energetically invested in the issue of traditional awards and worked hard to have them re-established. Rivers of ink and non-stop lobbying tried to influence the outcome of the Government's deliberations on the matter. In early 1980 a resolution of the issue was at hand; the Government resolved not to re-constitute traditional awards. The National Council changed tack and supported this, perhaps reading the writing on the wall. Major General Maitland had, at the PM's request, conducted a tri-service survey in Victoria re traditional and proposed awards. The outcome, from 'a questionnaire loaded for such a result in that it made it appear that the resumption of traditional awards was all but impossible'—agreed with the PM's views.[93]

For those who had so strongly supported the retention, however, it was a blow from their

National Council. It was discovered that the National Council had resolved to support the measures without a quorum as only NSW members attended the meeting that decided to endorse the Government's decision. It was another embarrassment to the National Council which had to reconvene a new meeting. The Victorian Committee was again irritated by the NSW-centric National Council and one member remarked 'that State branches should nominate two or three Councillors to the National Council to ensure that all of the decision-making and power to do things do not reside with NSW members [alone]'.[94]

The Victorians felt that they had not only been let down by the National Council, but deliberately sidelined on the issue. Their suspicions of the National Council would be amplified in August that year. Leading up to that was a discussion in the May committee meeting regarding the lack of a policy around universal military training. Bullard believed:

> ...the unit effort required for recruiting was about equal to the administration of an extra company. Additional ARA staff were needed to cope with this increased admin and there was also a requirement for AIC type staff [Australian Instructional Corps] to assist with unit training....the voluntary system had been used for about 15 years in the CMF and simply had not worked. Under this system, the right calibre was just not there.

Fellow committee member Brigadier Rossi agreed: 'it was evident that the voluntary system had never been sufficiently viable to raise a good officer corps and train sufficient numbers of NCOs to the required standards'. He maintained that universal training was the only method which produces a cost benefit solution to Australian defence.[95] It was expected that the Victorian motion would be discussed by the National Council in August.

In August 1980 the CMFA's AGM was held in Sydney. It was followed by a reception for the PM, Malcolm Fraser, attended by the CGS and major–generals of both the ARA and ARes and equivalent ranks in the RAAF.[96] The PM addressed the assembled Reserve and Regular officers there, emphasising the importance of the role of the CMFA. He said 'there was ample evidence that its representatives were being heard and he saw the organisation as the advocate of the Reserve Forces'. Fraser made it clear that the Association should enjoy the full cooperation of the regular forces.[97] This was indeed music to the CMFA ears, and those Reserve officers who attended that night would have left the reception feeling that their hard work over the past decade had been recognised at last.

The 1980 AGM was attended, in Victoria's case, by its vice-president Major General Poke who subsequently reported proceedings to the Victorian committee sub-branch at its committee meeting of 11 September. That committee, which had earlier resolved to put forward a motion for debate at the meeting following the national AGM with the aim of establishing a CMFA policy for universal military training, was disappointed to learn that not only had the motion not been put forward, but it also wasn't even aware that the meeting had taken place.

Red-faced, the state secretary Lieutenant Colonel Macleod, 'apologised for the lack of prior information regarding the National Council meeting but said he was quite unaware that it was to be held during August. It was most unfortunate that members had no opportunity to place matters on the agenda or to brief their representative [Poke] so that he could properly represent the Victorian branch'. One member declared that '...the lack of an agenda and the lack of briefing of our representative was unacceptable'. Another said that 'It appeared that the majority of the voting power rested with NSW'.[98]

Matters came to a head in December 1980. Invited to attend the December meeting of the Victorian committee were CMFA national president Major General John Macdonald and Colonel Glenny, the national secretary. After civilities and formal updates, the gloves came off. Lowen stated:

> ... as far as he was concerned the battle for the resumption of traditional awards had been won. In spite of the smoke screen put up by the bureaucracy, it had only needed the support of the National Council to bring it about, but unfortunately this had not been forthcoming. He had felt badly let down by this lack of support but took the view that we had at least accomplished the institution of a separate and unique system of awards for voluntary service apart from the National Medal. He appreciated that Major General Maitland, as a serving officer, had to follow instructions. This was not true of the National Council.[99]

Another member expressed his concern that the National Council 'continues to make decisions on important matters without prior reference to State branches. ...it was vitally important that the National Council had a constitution which was clear, workable and acceptable to all who worked with it...' Glenny tried to mollify the Victorians by stating that 'the National Council saw Victoria as the most active State branch followed by the Northern Territory'.[100] Macdonald's comments were not recorded.

The Victorians pushed on, one committee member saying, 'that it was vitally important that the national Council have a constitution which was clear workable and acceptable to all who were governed by it'. He offered to review the existing constitution, and this was agreed to. McLeod also complained about being 'denied the opportunity' to contribute to the previous year's agenda for the National Council meeting. Glenny admitted that pressures of time had beaten him, but he would now allow about three months to allow States time to work with him on the agenda.[101]

By April 1981 the review was underway, with the main aim 'to ensure that all states were fully represented on the National Council, that is, to make the Association truly national.'[102]

By the July meeting of the CMFA Victorian branch committee, the review of the CMFA constitution had been sent to the national president Major General Macdonald and the state president, Brigadier Lowen. It was meant to be tabled at the CMFA AGM scheduled for 24 August 1981 in Sydney. Macdonald was not happy; he 'expressed fears that the proposed revision was impractical, that adoption would lower the prestige of the council and that life members, from whom the Association derived political influence, would lose interest'. The Victorians, still

RANR personnel preparing to depart for the USA for training, 6 August 1981.
From L: AB Robert Healy, LS Dennis Mead, AB Bill Pollard, AB Stephen Wiggins, SMN John Sandow and AB Gary Cawthorn.
(Sea Power Centre—Australia)

smarting from their isolation from national CMFA affairs the previous year, were undeterred: 'It was agreed that the aim of making the Council a truly national body be maintained...'.[103]

The CMFA AGM was preceded on 20 August 1981 by a special meeting to discuss the new draft constitution and get everyone on the same page before the item went to the floor. At that meeting was Victorian State President Brigadier Lowen, National President Major General Macdonald and immediate past president Major General Murchison. '...to that date the only response had come from Major General Cullen who agreed with the 'National' concept but expressed concern regarding the difficulties of obtaining a quorum at meetings, the limitation on spending, and expressed the wish to retain distinguished life members on Council'. According to Lowen's later report to the Victorian branch committee, 'views were frankly exchanged'.[104] The outcome was that the National Council was to provide general guidelines and a sub-committee then formed to complete the draft within those guidelines.

Guidons, banners and colours of regiments took pride of place at Sydney's historic Victoria Barracks when 1200 male and female Army Reserve soldiers paraded for the rebirth of 2 Division and its return to the Army's Order of Battle in August 1981.
(Defence Images)

It seemed the Victorian branch committee had stirred the National Council into action. Stung by the quick action by the Victorian committee to draft a new constitution which was not at all to its liking, the National Council moved equally quickly to bring the process under its control. At the subsequent CMFA AGM, the guidelines were agreed (guidelines which should have been issued before the Victorians acted):

- That the National Council should be national in appearance and in fact so that no one state can control the council.

- The practice of appointing life members to the council will continue and be extended immediately by inviting a prominent citizen from each state: in NSW—Sir Roden Cutler; in Victoria—Rupert J Hamer; in SA, Sir Arthur Lee, in Tasmania Sir Harry Strutt, in Queensland, Sir Albert Abbott; with WA and NT to be nominated.

- Provision be made for a deputy or Senior Vice-President

- Provision for a quorum at meetings and appointments of proxies should be at a workable level (Victoria's Lieutenant Colonel Bullard had previously suggested the quorum should remain at 12 but if proxies not submitted proxy should automatically be the Chairman)

- No objection to immediate past CRES becoming Chairman (provided he is a member); it should not be an automatic appointment.
- Any financial restrictions placed on the National Executive shall be at such a level as will enable it to carry on normal activities without reference to the full council.

Membership of a sub-committee to redraft the constitution was also agreed—the CMFA national president or his deputy, along with Lieutenant General Brogan, and in a tilt to Victoria, Brigadier Lowen.[105] 'Brogan's draft' was to be ready by December.[106] Glenny would later comment in an interview that:

> ...there was a tension between the New South Wales, National DRA, and Victoria, and it was in a sense power influenced. Victoria saw itself as being low down the pecking order, and, not having the Cullens or the McDonalds or ...some of those people, and therefore subservient to the national body as such. There was certainly even then a degree of that when I arrived as Commander 3 Division. I was a New South Wales man posted in, why wasn't I a Victorian?[107]

The Constitution issue was not finally resolved until 1983 when Cullen, as actively involved as ever, informed Lowen that 'he applauds the latest draft of the constitution and will support its adoption' at the AGM in Canberra on 5 August 1983.[108] One wonders whether it was Cullen who insisted on that pre-meeting to discuss the constitution in 1981. In any case the new constitution was duly accepted.

The CMFA, formed in 1970, was despite these frictions finally moving to a constitution which was seen as equitable to all concerned almost 11 years later. The NSW 'establishment' would continue to be 'first among equals' when it came to influence in the Association but the progress in this matter alone showed the maturing of the Association. Frictions between branches and national bodies of associations were inevitable. What stood out with this association was the willingness, despite the professional jealousies and the personal differences, the state rivalries and the shocks of political changes and reviews affecting the CMF, that the officers involved remained committed to the betterment of the members of the CMF and to the survival of the CMF as an integral and time-honoured part of national defence.

3.
Towards a National Association

The National Council elected at the AGM in August 1981 saw further evolution and continuity, with a new president and treasurer. Life Members continued to play a role both symbolic and actual as their indirect and informal counsel was always valued:

    Major General Sir Victor Windeyer

    Major General Sir Denzil Macarthur-Onslow

    Major General Sir Ivan Dougherty

The National Council consisted of Councillors:

    President: Major General Macdonald

    Major General Cullen (Foundation President)

    Major General Murchison (Immediate Past President)

    Lieutenant General Sir Mervyn Brogan

    Brigadier Sir John Pagan

    Sir Colin Hines OBE [State President RSL]

    Colonel R. Garland

    Secretary: Colonel W. Glenny

    Treasurer: Brigadier Dart viz Lieutenant Colonel R. Faulks.

    Publisher *Citizen Soldier*: Lieutenant Colonel Nebenzhal/Major Crook.[109]

State branches were constituted as follows, with their presidents as vice-presidents of the National Council:

New South Wales–Colonel T Crawford

Victoria–Brigadier I Lowen

South Australia–Brigadier John G. McKinna CMG CBE DSO MVO KStJ

Western Australia–Colonel Robert D. Mercer AM RFD ED

Tasmania–Colonel L. Simpson

Queensland–Brigadier Kevin D. Whiting AM ED [110]

Northern Territory–Major J. Dunn [111]

At the 1981 AGM all states except Tasmania were represented, and guests included the Minister for Administrative Services, Kevin E. Newman MHR (Liberal/Bass)—a former ARA officer—Alan Edwards, the Chair of the CESRF (established in 1977); and Brigadier Colin N. Kahn DSO, representing CRES (Major General Maitland).[112] However, it was noted by Kahn that 9,000 enlistments per annum were needed to maintain a strength of 30,000—along with the regular cadre, equipment and 'maintenance of interesting training and sense of purpose.'

In some respects, the more important news was the announcement by Newman that legislation to implement supplementary compensation provisions for Reserve Force members, basically at the same level as for regular force members, would be enacted during the spring session of parliament. He stated that cover would be provided within the framework of the *Compensation (Commonwealth Employees) Act* and would provide for payments up to $500 per week.[113] The CMFA and its branches had lobbied hard for these provisions for several years. However, not all was as it seemed.

Later, Major General Kevin G. Cooke, RFD ED, Commander of 3 Division in Victoria, noted at a Victorian branch committee meeting:

> ...there were flaws in the amendments. The upper limit of $500 would not apply to C'wealth employees and the amounts paid would be reduced by such sums as were received from other sources. The members would be effectively subsidising the Govt to the extent of the other payments. This was seen as discriminatory and therefore unacceptable. Minister evasive on whether legislation would be retrospective.[114]

The Victorians agreed to approach Commander Graham Harris RFD, RANR to see if they could be kept informed of any developments. Harris was an MHR (Liberal/Chisholm) but it is not known whether he was indeed able to do so.

By early 1982 it was clear that the Army Reserve had made considerable progress, especially over the past three years. It expanded its strength by almost 40% and re-raised

2nd and 3rd Division Field Force Group. Reserve pay had been computerised and compensation had been introduced for reserve members. The Reserve Staff and Command Staff College had been established, and units had been established in the north of Australia—North-West Mobile Force (NORFORCE) and 5 Independent Rifle Company. A major works program had begun along with an increase in courses for Reserves and more and better clothing and equipment. The Reserve was employed on full time duty and had significant involvement in major Regular Army exercises. Recruiting funds and programs had been stepped up, with enlistment of specialists at appropriate ranks along with re-establishment of traditional unit titles and awards for long service.[115],

How much the CMFA did to affect any of this progress is certainly hard to quantify. A combination of CRES as a Reservist, the accumulated momentum of years of hard work by the National Council of the Association and all of the active branches through submissions, combining effectively with like-minded organisations such as the RSL and the Australian Defence Association (ADA), direct lobbying both formal and informal, and personal contacts and connections were all brought to bear. No one submission or lobbying effort could be pointed at as the tipping point for any particular campaign, but over this period some fruit was borne on the CMF tree.

RAAF reservist at No 22 Squadron at work, 1981
(Air Force Image Archive Collection)

As an indication of the *laissez faire* approach to funding and as well as the lack of close relations with branches, in March 1982 the National Council 'expressed disappointment' that the Victorian branch had not made any contribution to its funds. The Victorian branch committee then agreed on a 'gratuitous contribution' of $250 to the National Council. The committee noted that once the new constitution was adopted it would be bound to make payments to the National Council but at least the matter of contributions would 'be on a rational basis'. This was a revealing insight into the play between state branches and the National Council. If it were the same in other branches, little wonder that the National Council cum NSW branch was *laissez faire* itself in the way it consulted with the branches.

Lieutenant Laurie Wilson on HMAS *Bayonet* 1982.
(Sea Power Centre—Australia)

In early 1982, the 'Katter Committee'—named after C. Robert 'Bob' Katter Snr. MHR (Country Party/Kennedy) who chaired the Joint Standing Committee on Foreign Affairs Defence and Trade—announced that it would examine the structure of the Australian Defence Forces.[116] The CMFA and branches independently, including ever-energetic Victoria, decided to make submissions. Branches saw no conflict of interest in making their own submissions on issues often to the highest levels of Government, such as Lowen's letters to the PM re 5/6 RVR—'couched in forthright language that could not fail to be understood'—without National CMFA endorsement.[117]

Even though that was a particularly Victorian concern, nonetheless it reached across similar concerns in other jurisdictions. Equally possible was that the National Council at the time did not simply have the resources to chase every issue, with or without the branches' support. The lack of clerical, research and administrative resources would forever bedevil the National Council and most branches, limiting its ability to respond in depth and especially in a timely way to different concerns as they arose.

On 31 March 1982, Major General Maitland retired as CRES and was replaced by Major General Kevin R. Murray AO OBE ED QC (a Sydney Queen's Counsel and the first CMF general to reach that rank post-World War II) who in turn was replaced as Commander 2 Division by Brigadier Raymond ('Ray') J. Sharp RFD ED. These officers were all from NSW. While the succession plans rolled over, the CMFA started yet another year with some long-standing issues unresolved. That April, in *Citizen Soldier*, the President of the Association Major General Macdonald said:

> This Association must...continue to bring to the notice of the Government the need for effective callout legislation and to keep under review [the] introduction of improved conditions of service. The provision of home finance as a retention incentive will be raised again this year by this Association.[118]

The CMFA AGM and National Council meeting was held on 4 June 1982, once again at the Imperial Services Club in Sydney. States represented included NSW, Victoria, Tasmania, and the Northern Territory. Among the guests, apart from the CGS Lieutenant General Phillip H. Bennett CBE DSO, CRES Major General Murray and Commander 2 Division Major General Ray Sharp, were the Chairman CESRF Alan Edwards and the Minister for Defence Ian Sinclair (Country Party/New England). The President reported on the range of matters of interest and concern to the National Council at that time:

- CMF ceiling of the now achieved 30,000 and how to sustain it, let alone increase it.
- Call Out Legislation and the priority that the CMFA must give it.
- Adequate compensation for Reserve Forces.
- Defence service awards and the delays still being experienced in implementation.
- Defence Force housing loans for reservists.
- Death of Foundation Life Member Lieutenant General Sir Edmund Herring.
- The 'excellent progress' made by the NT, Tasmanian and WA branches of the CMFA.[119]

State reports included from NSW—'now an established branch'; Tasmania 'by far the most progressive branch', with 280 individual financial members and 12 corporate members; South Australia—Brigadier McKinna was seeking 'an active person' to become president; WA—membership at 125 and growing; NT—currently has 42 members 'although distance presents difficulties in communications and contact'; and Queensland—where the president was overseas but the branch 'seeking wider membership and increasing branch activities.'[120] It was a typical report at any time of the Association's development and evolution into a truly national organisation. Some branches were doing exceedingly well and making a real contribution to both public and CMF awareness of the Association and its works. Others were struggling to find the right mix of leadership and membership.

Alan Edwards, Chairman of the CESRF addressed the AGM, noting that in NSW all but 400 of some 8000 reservists indicated that they belonged to organisations covered by CESRF. While organisations represented in the CESRF covered 750,000 employees in Victoria and '87%' of the private work force in South Australia, Edwards gave a general briefing covering matters relating to employer support, the future of CESRF and problems facing the CESRF and employers.

Problems faced included employers trying to meet requests for leave beyond the 16 days contract—up to four weeks in some cases. When added to the four-five standard weeks of leave, this certainly imposed strains and burdens on employers. Edwards also noted that most employers believe the Reserves can be called out and support them because of this. The Government, said Edwards, must introduce Call Out Legislation to enhance Reserve credibility.

In the audience was John Stetson of the US Employer Committee of National Guard and Reserves, who stated that 'while the CESRF started four years later than his own organisation the CESRF is also making faster and more significant progress'. Yet, dealing as it was with the commercial enterprises within its membership, the CESRF had to tread a fine line between the push for greater support for the reservists working in those enterprises with the conservative business driven needs of those businesses. Despite the reports and the positive sounding information, the old dictum of 'make haste slowly' certainly applied to the CESRF, although Edwards estimated that '1200-1500 reservists remained in the forces as a result of CESRF efforts.'[121] It is not clear whether this claim was substantiated at the AGM.

After Edwards came the CGS, Lieutenant General Bennett, who 'acknowledged the role the Association had in representing the defence needs of Australia to the Government'.

ABRO Bob Healy, RANR on HMAS *Ardent*, 1982.
(Sea Power Centre—Australia)

It was curious that Bennett did not say 'represent the needs of Reserves in the defence of Australia', but his mere presence (along with the Minister for Defence) underlined the Association's growing status at the time. The CGS also acknowledged the work of the CESRF and CMFA in making the expansion phase of the CMF strength 'an outstanding success', despite some issues arising. His general briefing on the current and future Army concluded with 'an assurance that any initiatives introduced would not lead to any further major restructuring or reorganisation of the Army Reserve.'[122]

Bennett, soon to be knighted in the 1983 New Year Honours List, made observations on the objectives of the Army and the problems in achieving them but also included an appeal to the CMFA for its support 'to ensure that the Reserve understands the these problems, but that positive actions are being taken.'[123] Finally, the Minister for Defence Ian Sinclair addressed the meeting luncheon, also noting the value that he and the Government placed on the CMFA and its role including in maintaining defence awareness and keeping the Government informed.[124] All heady stuff for the National Council, although Major General Cullen if present, would no doubt have been unimpressed, yet pleased that the Association that he had formed would win such public accolades from such public figures.

After lunch topics of the discussion included Defence Force awards, Call Out Legislation with a Victorian submission forwarded to the CDFS and CGS, which had been used by the CMFA as the basis for a letter to the Minister for Defence on the subject. The need for key civilian employers to be briefed by CESRF regarding support for the legislation and Defence Force housing loans for reservists with a submission given to Councillors and Minister and CRES briefed was also discussed. As well, community awareness; defence restrictions on attendance allowance and hence training; and the Constitution which 'gave rise to much discussion but no decision' were included. Other topics discussed included Repatriation benefits for members of UN Observer Teams and peace-keeping forces and appointment of life members.

The election of a new National Executive followed. Major General Maitland became the new president vice Major General Macdonald, while Colonel Glenny as secretary and Brigadier Dart as treasurer were both confirmed. Lieutenant Colonel Nebenzahl, who had been at the original formation of the Association was once again appointed editor of *Citizen Soldier*.[125]

Again, the political winds were shifting. In March 1983 the Labor Party swept back to power under PM Robert ('Bob') J. L. Hawke (Labor/Wills). That a new and divisive issue would arise as a consequence six months later could not have been anticipated by the CMFA. It continued to pursue the Call Out Legislation issue which, with the change of Government went back to the start for further review by Government.

Submissions to the Sub-Committee on Defence Matters of the Parliamentary Joint Committee on Foreign Affairs and Defence were being completed by the CMFA and branches (apparently in isolation from each other). With Gordon G. D. Scholes (Labor/Corio) now the Defence Minister, at least one submission was sent directly to Scholes. In June the Victorians in their cover letter to that submission to Scholes, noted: 'In view of the changed circumstances, and the reviews you have instituted as Minister, we believe it proper that this submission should be sent directly to you.'[126]

At the National Council AGM and meeting on 5 August 1983—held at the 3 RNSWR training depot in Canberra—many of the usual issues were considered. As earlier described, the new Constitution was finally adopted. Reports were given noting no progress on the issues of granting Defence Service Home loans to reservists or on Call Out Legislation. A proposal to integrate Regular and Reserve units was also discussed (no decision had been made by Army).

Soldiers of Support Company, 4 Battalion, RNSW Regiment, prepare to storm ashore at Birkin Head Point in September 1983. The mock assault was a highlight recently of a two day Army reserve promotion at the shopping centre.
(Defence Images)

Major General Cullen submitted a table showing the percentage of former regular officers at lieutenant colonel level and above, in ARes units in 2 Military District (2 MD). He noted that it was entirely up to CMF commanders as to whether or not they accept ex-regular officers into the Reserve. The issue of long service awards was finally ending with awards to begin in 1984. Major General Maitland was re-elected CMFA president.[127]

Soon after came the bombshell from the Government—the 1983/84 Budget saw a decision to tax Reservist pay. This decision clearly and adversely affecting the Reserves, such as training time, strength reductions including of cadre staff and even support for cadets. The tax issue in particular immediately exercised the National Council and all State branches. The backlash put the Government on the back foot, and it then referred the decision to the Committee of Reference for Defence Force Pay (the 'Coldham Committee'). It was an issue that united the CMFA like never before.

Special meetings were held in CMFA branches and the National Council in early September 1983 to discuss the effects of the Government budget decision to tax Reserve pay. Reserve pay had been untaxed since 1964.

Tasmania and Victoria also sent representatives to the national meeting in Sydney while WA and NT made submissions. It was agreed that each branch should make its own submissions to the Coldham Committee directly with the first priority to have the tax decision reversed.

The 'fall back' position was that pay and allowances to be adjusted to at least the same level as the regulars on a working day basis; and taxation should be at a flat rate. The president and delegation would seek an audience with the PM to 'convince [him] that the decision would do irreparable damage to the ARes with little or no financial gain to the Government'. *Citizen Soldier* would inform members of what the Association was doing and encourage them to make submissions to the Coldham Committee. The Coldham Committee met with Reserve officers around Australia; invariably this included RANR and RAAF Reserve officers as well as ARes.[128]

Through 1984 the CMFA was busy on many fronts. It especially continued the pressure in support of Justice Coldham's imminent recommendations re taxation of ARes pay. State branches reached out to allied organisations for support. For example, Ian Spicer, the Executive Director of the Victorian Employers Federation, made a submission opposing taxation of Reserve forces to the Minister of Defence. Meanwhile, the Chairman, CESRF also wrote to the Minister re taxation. The letter included a paragraph pointing out that while taxing reserve forces pay would marginally increase revenue, the expected pay rises [to compensate for the tax] will come from the Defence vote, leaving less for important defence items.[129]

Changes were afoot in the ARes. Major General Murray's term as CRES was extended to March 1985, but Reserve Branch at the Department of Defence was closed at the same time and replaced by the Office of the Chief of Reserves, a substantial downgrade in status, according to many. Senior regular staff were re-posted, and only clerical type assistance was provided. In an ironic change to Reserve unit efficiency, some units were overstrength and other units under. Not so long ago, units with less than 70% strength were disbanded; now they were being forced to have less. Meanwhile no unit was allowed to expand and the Government restricted unit ceilings and reduced training days.

Meanwhile routine matters remained to be implemented—late in 1983 the National Council accepted Major General Sir Robert Risson as a life member, followed by Colonel Sir Bernard J. Callinan CBE DSO MC and Colonel Sir John G. Norris KC ED, in early 1984 (all Victorians). The Association also took up the concern that the Army's traditional championship medal for rifle shooting, the Queen's Medal, was to be scrapped. Curiously, despite the apparent solidarity among the CMFA organisations regarding the tax issue, friction still existed. Lowen stated to his Victorian committee (in words he would later regret) that:

> ... he was strongly opposed to the situation that obtains in NSW, due to the proximity of the National Council, developing in any other state. The state branches must flourish and not be over-shadowed by the national Council if co-located.[130]

Meanwhile the Tasmanian president Colonel Simpson had aligned Senator R. Brian W. Harradine (Independent/Tasmania) with CMFA views and try to have him raise the issue in the Senate, although Harradine said that he did not think the Liberals would do anything to prevent the imposition of the tax.[131] The CMFA had stronger allies in parliament, such as Kenneth ('Ken') J. Aldred MHR (Liberal/Bruce). Aldred supported CMFA positions on several matters. For example in March 1984 he wrote to the now Shadow Minister for Defence Ian Sinclair MP about Call Out Legislation.[132]

On 31 May 1984, Aldred made a statement in Parliament during its Grievance Debate.[133] Hansard recorded the content of that statement. As a robust statement in support of the ARes and the CMFA it has been shown here in full:

> **Mr ALDRED** (Bruce) (1.25)-My grievance today, I am sure, is shared by the greater majority of this nation's people. It is a very real concern over the Hawke Government's persistent, continued and unrelenting determination to undermine Australia's defences. As usual, whenever a Labor government moves against the defence forces, it is the reserves that bear much of the brunt, especially the Army Reserve. This Government started to emasculate the Army Reserve in the last Budget and has continued its destruction since then. Before the Minister for Defence (Mr Scholes) discounts this claim as an exaggeration of mine, let me tell this House that this feeling is shared not only by members of the Reserve but also by the Citizen Military Forces Association and senior regular Army officers.
>
> Concern is so great among members of the Victorian branch of the CMF Association that its President, Brigadier Ian Lowen, on 21 May this year, wrote to the Prime Minister (Mr Hawke) expressing the Association's concern. In that letter he told the Prime Minister that members of the CMF Association were 'increasingly concerned about current trends in the Army Reserve'. Men of the standing of Brigadier Lowen do not easily write to Prime Ministers with complaints about their policies. In fact, the Brigadier wrote on behalf of his organisation to the Government twice in September, then in February and again in May twice. What grievances do these men and women of the Reserve have? These men, many of whom have devoted many years' service, are distressed to see the Army Reserve disintegrating because of this Government's actions.
>
> Let us look at some salient figures. In the first four months of this year Army Reserve officer resignations and transfers to the inactive list were up 70 per cent on those for the same periods in 1981, 1982 and 1983. What is even more disturbing is that the majority of the reservists leaving are in the middle ranks, that is, senior non-commissioned officers and officers with skills in medicine, other professions and in technical trades. It is the high achievers who are being driven out of the reserves. It has taken many years to train these key men and their leaving will have a cumulative effect on the regular services also. There are several reasons for this exodus, all of which have been caused by this Government's changes in policy towards the Reserve. The erratic policy of taxing

reservists' pay, later to be changed by the besieged Defence Minister to partial taxing, has been a major blow to reservists.

As the official journal of the CMF Association, *Citizen Soldier*, writes:

> The Government expects the CMF to continue to work unchanged out of a spirit of patriotism—Mr Hawke said so. Mr Hawke and his Cabinet haven't taken a half cut in salary—do they lack patriotism for this country of ours? Taking the kindest view, the Government took its steps without realisation of their effect. It certainly took steps without even seeking, let alone obtaining military advice. Indeed, here is a Labor Government which professes its intent to communicate disregarding any form of consultation, even with the Government's own committee for that purpose—the Army Reserve Advisory Committee.

That is what the CMF Association's journal thinks of the Government. It is now a matter of record that the Government announced the taxation measures without any mechanism in mind. The consequences of this decision have been horrific for many reservists. For instance, the ordinary working man serving in the Reserve has been caught up in the payment of provisional tax for the first time. The members of university regiments were also hard hit by this decision. Let me quote the words of Mrs J. Millett, from Belmont, which adjoins the electorate of the Minister. In a letter to the *Melbourne Herald* of 24 October last year Mrs Millett told us that because her son, a member of the Deakin University Unit, now had to pay tax he will no longer be eligible for the tertiary education assistance scheme allowance for students. Mrs Millett also did some quick calculations regarding her husband's situation, who is and has been a member of the Reserve for the last 35 years. Her calculations show that for the privilege of being a member of the Army Reserve Mrs Millett will be paying out $1.46 per week. Mrs Millett concludes:

> The Government should have had the fortitude to disband the Army Reserve in the first place if this is the ultimate intention, instead of resorting to this type of subterfuge.

Mrs Millett's concern is widely shared. Some 3,000 submissions have been made to the Coldham Committee, which is currently reviewing Reserve pay. The overwhelming majority of these have vigorously opposed the removal of tax-exempt status for Reserve pay. This Government may be able to fool the public for some time, but it will be caught out. On this occasion the Minister and the Hawke Government have been caught out in their determination to destroy the Army Reserve.

It was this Government that announced last year cuts in the number of annual training days for reservists from 38.25 days to 36 days. In reality senior officers report that training days have actually been cut to as few as 30 days and, for some units, even fewer. If the Defence Minister takes off his tin hat and listens to his Army Reserve advisers, he will learn that for a soldier to have the bare minimum of efficiency he needs at least 26 days of training. In fact, senior Reserve officers regard the desirable levels of training for field forces as 43.3 days, for logistics command 45 days and for training command 52 days. But what is happening? All around Australia recruit courses have been cancelled with two recruit courses cancelled in Victoria so far this year. The reason? Lack of man days and insufficient recruits. The military skills competition for the Third Military District, scheduled for November last year, was cancelled, again through lack of man days. The 2nd Battalion, Royal Victoria Regiment has not had a full parade since late 1983. The reason? Lack of man days.

But this Government continues its subterfuge even further. In August, the Minister for Defence said that the Reserve would be maintained at 30,000. A positive statement? No, a statement of deceit. The Army Reserve posted strength at that time was 33,523. The Minister was actually announcing a cut in Army reserves. The reserves have been ordered to reduce their numbers by 30 June this year. This is already under way with many units now being forced to discharge efficient soldiers. At a time when tension in the world is high and obviously set to heighten this Government cuts back on Australia's defences. In reality the reserves will be cut back to an effective force of around 24,000 to 25,000 or even less, as the 30,000-target announced by the Minister must also include reservists with less than the 26 days efficiency requirement. The field forces will be hardest hit. In my State of Victoria, the Army Reserve field force will be reduced by 1,200. This is a ludicrous situation. At the very least, administrative procedures should be simplified so that Reserve units can expeditiously discharge inefficient soldiers and replace them on posted strength with efficient soldiers as soon as possible.

Is it any wonder that morale is rock bottom and resignations are ever increasing? I know that the men and women of the Army Reserve are dedicated but there is just so much kicking from this Government that they can take and kicks they have got. Yet another kick came in February 1984, a few months ago when the Reserve Branch in the Army office was abolished. This step was taken in spite of contrary recommendations from the Millar Committee of Inquiry into the Citizen Military Forces to establish the Reserve Branch. Let us not forget that the Millar inquiry was set up by a Labor government. The time has come for this Government to come clean. It must cease hiding behind deceit, subterfuge, and false statements of support for the Army Reserve. This Government should announce in the coming Budget that it intends to abolish the Reserve. It should admit its true aims and then take the consequences from the Australian public.[134]

In August 1984, prior to the CMFA AGM scheduled for late that month, it was reported that the Minister for Defence had accepted the Coldham Committee recommendations on taxation of Defence Force Reserves pay, i.e. that a range of salary and attendance allowances to be introduced backdated to 1983. Doubts were expressed that the new allowances, fully taxable, would have any effect on retention. The Shadow Minister for Defence, Sinclair, had already promised Aldred that the Liberals would abolish the tax if returned to office. The Coldham Committee was to be replaced by a Remuneration Tribunal.

The 1984 CMFA AGM and meeting was held at the Imperial Service Club in Sydney on 31 August. On the agenda was business arising from previous minutes and ongoing such as housing loans, the bicentennial medal, integration of regular and reserve units, the future of HQ 3 Division, Call Out Legislation. The main agenda topics included force structure, including integration, pay and taxation, including attendance and Tertiary Education Assistance Scheme [TEAS] allowances; the submission the Defence Sub-Committee of the Parliamentary Standing Committee and senior ARes officer appointments.[135]

Major General Maitland announced he would not be available for re-election as President. In a remarkable change of direction Brigadier Lowen from Victoria was elected national president and Lieutenant Colonel Nebenzahl as deputy president. In a further big change to the Association, it would now be based in Victoria. Lieutenant Colonel D. MacLeod was appointed secretary and Colonel Donald ('Don') J. Sandow RFD ED as treasurer.[136] Council records were to be transferred to Victoria. There was some discussion of the need to 'rejuvenate' the NSW and SA branches. Later Lieutenant Colonel Bullard was elected president of the Victoria branch vice Lowen.

Brigadier Lowen had announced to the Victorian Committee in June that he did not intend to stand for re-election and the possibility of the move of the National Council was clearly known at least in Victoria. Lowen was recognised for his energy, willingness to lobby Canberra directly on a range of matters and for his success in having Reserve service awards accepted, The NSW branch and delegates seemed less effective; and Maitland for whatever the reason, had stepped back perhaps prematurely from the presidency.

Not surprisingly, just as the National Council had often appeared to work in the same envelope as the NSW branch of the Association, for which it had been criticised including by Lowen and the Victoria branch committee, the same happened in Victoria now that the National Council was resident there. Brigadier Lowen invariably attended the Victoria branch meetings, and Lieutenant Colonel MacLeod continued as membership secretary for the branch for some months as well as National Secretary for the Association.[137] Victoria branch minutes over the period sees a mixing of national and state matters. At least the Victorian branch was well informed and acted as an extended secretariat and sounding board for the Association's executive. By December Lowen was reporting that 'State branches appear to be becoming more active. The NSW branch has [Brigadier R. S. Philip ('Phil') Amos RFD ED] as president and is being re-activated'.[138] Perhaps the move from NSW of the National Council had indeed allowed, in this case, the NSW branch to 'flourish'.

At the 1984 meeting, CRES noted that man day restrictions were easing as numbers declined; total strength was down to about 26,000 and still going down, but 30,000 was the goal for average strength. Retention numbers were still declining. However, successes for the year included the establishment of the Reserve Command and Staff College, Prince of Wales Scholarships for overseas attachments of a limited number of reservists and overseas training of ARes contingents. Other discussion centred around the plans to integrate ARes units with regular. CMFA officers felt this could destroy the territorial characteristics and identity of the ARes and severely limit the progression of officers to senior command and staff appointments.[139] Alan Edwards, Chairman of the CESRF, also reported progress, with $500,000 put aside for recruiting among other matters.

The founder of the CMFA, Major General Cullen, also spoke at this AGM, proposing financial support from Government to establish a 'Defence Support Organisations Secretariat' (DSOS) whose purpose would be collate information to be made available to supporting organisations including the CMFA, as well as MPs, media and the public. Brigadier Lowen undertook to sound out Sir A. G. William ('Bill') Keys AC OBE MC, National President of the RSL, about the proposal.[140] Lowen later reported that attempts to enlist RSL support for the establishment in Canberra of a DSOS 'have met with no success at all. My feeling now is that we now have too many organisations concerned with defence and certainly one such as DSOS would have little chance of success without RSL support...'[141]

The newly established Defence Force Remuneration Tribunal (DFRT—replacing the Coldham Committee) would carry on the enquiry into pay and allowances. The Association decided to make a submission to the DFRT. A paper, supporting the concept of tax-free pay and removal of allowances, was subsequently also forwarded to the Treasurer Paul J. Keating (Labor/Blaxland) and Defence Minister Kim C. Beazley (Labor/Swan). It made several recommendations around rates of pay ('should be the same for ARes and Regulars') and Reserve Allowance ('remove it and return to full tax-free status or make the allowance tax-free.')[142] Other branches, such as Queensland, also made submissions.

Meanwhile, Brigadier Lowen reported to a Victorian branch meeting that National Country Party leader Ian Sinclair had responded to Association requests for Call Out Legislation to be progressed in the negative; he simply expressed disappointment at the failure by the Government to 'grasp the nettle'. He also noted that the [Paul] Dibb review of the Defence Forces commissioned by the Government in February 1985 might result in Army taking the biggest cuts.[143]

4.

The Total Army and the Dibb Report

Major General Cullen continued to be directly involved in any matter of detail affecting the Reserves. For example, in October 1984 he wrote to Brigadier Lowen as the new national president relating some segments of a speech by the then Defence Minister Scholes at a UNSWR regimental dinner and conversations he had with him soon after at a United Services Institute (USI) function. Scholes had made some statements about ARes generals being unlikely to command in any future war which Cullen queried with Scholes, who replied that 'with the proposed integration for the ARes by units etc. with regular units I doubt if there will be ARes formations for them to command'. Cullen wrote: 'The inference is that this could be quite soon in the future. I assumed he meant divisional formations, but just what did he mean?'.

Cullen went on to say:

> Anyway, I regard it as very disturbing—especially the implications.
> And what exactly has he been told, and by whom?
> Is Mr. Mills the source? (his principal private Secretary—Mr. Mills was quite cross with me!!!)
> Is it [Lieutenant General] Sir Philip Bennett?
> Is it [Lieutenant General] Peter Gration?
> Or who?
> And what do they intend?
> And what do they really mean?
> And what should we do about it?
> It is clear that more information is needed about both the thoughts and reports of certain people, and the facts.
> All unclear, but of great concern
> We must talk and act about it.[144]

No doubt Lowen was used to Cullen's approach to these matters, but Cullen had learned long ago to trust his instincts when it came to hidden messages in statements seemingly casually made, by Defence Ministers. And even now, 15 years after the establishment of the CMFA, Cullen was not about to let anyone off the hook if they were looking to attack his beloved CMF. Cullen's approach was wholeheartedly endorsed by the NSW branch, which also wrote to both Lowen and the Defence Minister 'expressing...concern at both the idea of integration together with lack of information on such a proposal....'[145]

Correspondence from NSW branch president Brigadier Amos to national president Brigadier Lowen and return was regular as they exchanged views on a range of matters. For example, about Honours and Awards, i.e. the apparent discrimination against ARes members in terms of numbers of awards compared to the ARA. This was described in one letter from Lowen as 'a hardy annual' which he had investigated over five years earlier while in the Victorian branch committee. But he had to 'eat humble pie' when it became clear that the trouble then lay in the quality and number of CMF submissions. He questioned whether in NSW this was still the case?

Another perennial issue was the transfer of ARA officers to the ARes. Here Lowen said:

> With the new Force structure, increased ARA/ARES integration, and changing role of the ARES, all of which have some merit, we must ensure that the ARES does not lose its identity, lose units and traditions, and end up with all senior commanders from unit level upwards being regular officers...otherwise all the best officers will quit at major level.[146]

The ARes case for promotions and appointments to command positions was not helped by the results of the NSW senior TAC5 Course (for promotion to lieutenant colonel) at the Land Warfare Centre in Queensland in 1984—of 34 participants, 28 failed the course. This has 'thrown a dismal blanket over the Reserve, and of course, invite more ex-Regulars to join in the absence of qualified reservists.'[147]

Lowen had also initiated, in the interests of better communication, a series of newsletters to all members of the National Council and State delegates, with four having been issued in 1985 to date, and another due out in November.[148] The June 1985 newsletter for example, noted CMFA submissions (specifically from Queensland branch) to the DFRT had been made on rates of pay and attendance allowances. Another reason to move to newsletters might have been financial—the auditor's report for June 1985 showed the Association to be nearly $4000 in the red, with the major part of that being the cost to produce *Citizen Soldier*.[149]

Another submission was made to the Senate Standing Committee on Foreign Affairs & Defence on the Army's rapid deployment capability and with particular reference to the support roles and expansion base aspects involving the Army Reserve. The submission made no less than 16 recommendations.[150] The CMFA had also provided Paul Dibb with a copy of the submissions on the 'Structure of the Australian Army' (a Victorian branch paper), as well as the paper on rapid deployment—the Dibb Report was due to Government in March 1986.

In other advocacy, the Association president had once again written to both Minister for Defence and the Treasurer regarding taxation of Reserve pay. Minister for Defence Beazley had responded in January 1985 in a letter covering a range of matters including that the Government, *'having given special consideration on the views of Reserve Force members'* [author's italics], had reduced the taxed status of Reserve pay by 50%.[151] Beazley's reply did little to overcome the concern about that Government's attitude to Defence.

Lowen noted the political support given the situation in the Reserve by the introduction for discussion in the House of Representatives of the 'alarming run-down by the Government of the manpower, morale and effectiveness of the Army Reserve' by Ken Aldred MP, supported by Robert G. Halverston, OBE MP (Liberal/Casey; a former Air Force officer).[152]

Aldred's statement included the following:

> By the end of April 1985 the posted strength of the Reserve was down to 23,900.... in August 1983 when the first Hawke Labor Budget was brought down, the strength of the Army Reserve was 33,523...over the past 20 months ...we have lost 10,000 reservists. One-third of the Army Reserve has voted with its feet and left.... There is now a national shortfall of 1,000 officers. Over the past 12 months the average rate of officers and NCOs leaving the Reserve has doubled.... How did it happen? It happened because this Government removed the tax-free incentive granted to the reserves in 1964 by the Menzies Government.[153]

Referring to the discussion, Lowen noted the replies by the Government members 'as might be expected dealt with almost everything except the problem of the ARes'.[154] There was clearly a certain level of suspicion regarding the Labor Party's commitment to Defence, let alone the Reserves.

Meanwhile, in view of the forthcoming general election, Brigadier Lowen had asked for and quickly received, the opposition parties' defence and defence support policies. The Opposition had in turn asked for a CMFA submission regarding those policies—State branches were invited to send submissions to the CMFA executive so that collated submission could be made to the parties' defence committees. High on the list of concerns would be the Call Out Legislation, without which the proposed integration of Regular Army and Reserves will not work, and another round of the tax question.

The 1985 National Council AGM was held in Melbourne on 30 August. It was the first meeting to be held in Victoria. As usual after CMFA AGMs there would be a dinner for delegates and wives. For the first time in 15 years this would be held not at the Imperial Services Club in Sydney but at the Naval & Military Club in Melbourne, with its showcase collection demonstrating Victoria's long military tradition and contribution to Australia's defence.

The meeting agenda noted that the morning meeting with the CESRF would be addressed by the Minister assisting the Minister for Defence Michael J. Duffy MP QC (Labor/Holt) and the CGS, Lieutenant General Peter C. Gration AO MBE. Other guests to participate were CRES, Major General Cooke and Commander 3 Division, Major General James ('Jim') E. Barry MBE RFD ED.[155] One item discussed was the appropriateness of the present name of the Association and whether it properly reflected the change from CMF to Army Reserve.[156]

Later, updating the Victorian branch on the meeting, Lowen said that all state branches, with the exception of Northern Territory and South Australia, were doing well; in the NT, a previously strong membership area, a change of leadership had resulted in the CMFA losing touch to some extent.[157] By February 1986, Lowen reported in the national newsletter:

I regret to say that despite several letters I have yet to hear anything from this once active branch, and I'm beginning to wonder if it has been wiped out by another cyclone Tracy that we have not heard about down here. This would be a great pity, as, like Queensland, they are in the forefront of integration.'[158]

However, total CMFA membership was about 1000, and all State delegates were now appointed. At the meeting itself, the branches reported mixed results. The newly established Queensland branch now had about 60 members and a full committee. Brigadier Amos, the new NSW branch president, stated 'that when he took over the [NSW] branch had virtually ceased to exist, due to the proximity of the National Council. Now there was a full committee, meeting every six weeks, and a financial membership of about 20'. In fact, the stated purpose of the NSW branch meeting in October 1984 at which Amos was elected State president was 'to reform the [NSW] branch of the [CMFA]'. Its first step was to adopt the constitution of the national body.[159]

Victoria boasted 278 members and while they were fortunate to have the national president on their committee, they 'were careful not to duplicate the efforts of the National Council'. WA had about 60 members and capitation fees would be paid from 1 July 1985, implying that the branch was also just getting established, while South Australia did not attend the meeting at all and submitted no report, a reflection of the low level of activity in that state.[160]

Perhaps stung into action, the South Australia president Colonel Raymond J. Stanley CBE RFD ED KSJ reported in February 1986 that he intended, starting that month, to build the branch into 'an active, supportive and involved association.'[161]

Tasmania was the standout branch and its 582 members (478 financial), i.e. about half of the CMFA's total membership, really showed how moribund the NSW branch had become. NSW, home to the Reserve 'establishment' with its senior CMF coterie of general ranks and home to the premier Reserve 2 Division, really had fallen by the wayside. Even though this would change again under the determined leadership of Brigadier Amos, there is little doubt that the move of the National Council was a timely catalyst for renewal in NSW. On the other hand, how and why did Tasmania, in real terms perhaps the least 'endowed' with defence assets and regular let alone reserve units, come to be so large?

Tasmania branch was certainly active in different ways to the other branches. It was working with Senator Harradine to set up a housing loan scheme for its members; it sponsored a runner in the Sydney-Melbourne marathon; the president acted a parade host for the presentation of colours to the 12th and 40th Independent Companies; 'social activities were maintained at a high standard and included the highly successful annual dinner dance and family gatherings' while the branch was developing plans for junior NCO travel grants, unit liaison activities and is assisting in the setting up of a Tasmanian branch of the Australian Defence Association (ADA).[162]

There was no doubt that Tasmania with its long Reserve tradition and by acting locally, was gaining, and maintaining community support for the Reserves. It was one of the first branches to welcome Naval and Air Force Reserve officers into its ranks. Perhaps in recognition of Tasmania's position its President, Colonel Simpson was elected as CMFA Vice-President, replacing Lieutenant Colonel Nebenzahl.

Concerning like-minded organisations, Lowen reported that he had been approached by the Armed Forces Federation of Australia (AFFA) with a view to pursuing common aims. The CMFA however was primarily interested in the defence of Australia and the part the Army Reserve plays in it, while AFFA was primarily interested in pay and conditions—basic information exchange will be made. In another sign of the changing times, Lowen did note that AFFA had an Air Force officer as a Vice-President—Reserves.

Later, in mid-1986 the CMFA almost circumstantially became involved in a contretemps with the RSL over the Association's membership of the Australian Veterans and Defence Services Council (AVADSC), an umbrella for ex-service organisations. The CMFA, which had joined AVADSC as a member back in 1977 but paid no subscriptions there since 1979, was still listed by that organisation as a member when AVADSC came into dispute with the RSL; the RSL thought CMFA was part of their problem with AVADSC. Major General Murchison, who had been attending AVADSC on behalf of the CMFA would give a briefing at the 1986 AGM.

Returning to the 1985 AGM, Lowen noted that the CMFA had made a submission to Government about recognition for Australian service in the Sinai, and this had met with some success. He hoped that that success would rub off on the CMFA's submissions to Government re recognition of Vietnam service and of the Queen's Jubilee Medal for Army Reserve members who qualified. Award of the Vietnam Medal for ARes officers who had been on attached duty in SVN during that war was subsequently rejected by the Defence Minister Beazley.[163] Last Brigadier Lowen made a general statement regarding relations with Army Office and the standing of the CMFA:

> ...for some time the Executive had been trying to establish a link with Army Office, so that we are kept fully informed of what is currently going on and what is planned. This information will ensure that the Association's efforts are properly directed, and a bond of mutual confidence is established.
>
> After a short session with Lieutenant General Gration, he [Brigadier Lowen] was given a full briefing by the Chief of Operations [Brigadier Fisher], Director-General of Development [Brigadier Bird], Director-General of Army Training, the Director of Army and the Deputy CRES [Brigadier Hodges] on the whole scope of the re-development plan of the Australian Army, with particular emphasis on the Army Reserve. The briefing was very thorough and will prove most valuable in working out the Association's aims and activities.[164]

In his report as president for 1984/1985, referring to that meeting, Lowen said that 'It is gratifying that the CMFA is recognised as a responsible body, and considered worthy of being taken into the confidence of those responsible for planning the future of our Army Reserve.'[165] The president of the Victorian branch, Lieutenant Colonel Bullard, added to Lowen's statement that 'this year, probably for the first time, the Association was working on a truly national basis.'[166] These events illustrate that for the CMFA, relationships needed for advocacy, lobbying and information exchange were ever evolving within the Defence and political landscape. Clearly some state branches were doing well at different times, depending on the leadership and quality of their members and committees.[167] The same could be said of the National Council and its executive.

With Brigadier Lowen, it seems the CMFA was in good hands. An exchange between Lowen and the Minister for Defence is also illustrative of the response by Government to the Association—and Lowen's—influence. After yet another letter to the minister about tax exemption for ARes pay, the minister replied that this was not the only reason there was a fall in ARes numbers. So, the government had commissioned consultants to investigate; the study would take about four months. Lowen reported to the members: 'I have of course replied that he could obtain the answers more quickly, and free of cost from either CMFA, or from discussions with CRES'. While it is not clear whether Government had commissioned a study to sideline CMFA pressure, it would not have been surprising given Lowen's persistent and direct advocacy highlighting Government inaction.[168]

Taxation remained a major issue for the CMFA, and the 'new' NSW branch was soon to submit an entirely new approach to taxation. It planned to stop trying to return the tax exemption to reservists but rather push for full equivalence in pay and allowances with the Regulars. For example, conditions such as sick leave, annual (recreation) leave, long service leave and contributions to the Defence Forces Retirement and Death Benefits (DFRDB) scheme.[169] Opposed by a majority, the idea was quietly dropped and the status quo accepted—only a full return to tax free status would be acceptable.[170] By February 1986, integration was proceeding apace, with Queensland with its ARes strength second only to NSW, of particular interest with the anticipated disbandment of units, amalgamation of others and especially the commands going to regular officers.

In December 1985, Lowen reported that he had been informed by CRES, Major General Cooke, that the current strength of the Army Reserves was 23,673, well below the Government's stated position of 30,000 average strength. By February 1986 strength was 23,440 but included 3000 non-effectives. From 1 July 85 to 31 October 1985 recruitments had been 2685 but discharges 2935, a net loss of 250.[171] This was of major concern.

Despite a recruiting drive and advertising campaign the situation was not improving. Overshadowing this concern was the anticipated release of the Dibb report. Lowen noted in one of his newsletters:

> It is significant that perhaps that the CGS has considered it necessary to assure the Army that it has 'nothing to fear' from the report. As Dibb originally stated that he thought any changes he might suggest could be achieved without increasing the Defence Budget it seems that someone must suffer. I wonder where the axe will fall?[172]

As late as February 1986 the NSW branch was asking for a copy of the 'official Army policy to integration.'[173] However, integration of Regular Army and the Reserve was already essentially in place by April 1986 and the CMFA essentially stopped campaigning against it. Even the CMFA NSW branch stated in its June 1986 newsletter: 'The 'one-force' concept has inescapable logic but must be genuine integration and not the use of the reserve as ARA cannon fodder.'[174] Perhaps accepting the inevitable was premature given what was to come, as shall be seen, with a Defence White Paper due for release later that year.[175]

It was the start of the 'Total Force' concept and in theory at least gave a role to an upgraded Reserve as part of the operational first line, with other units having a key base expansion role. Many were hopeful that the Reserve would emerge with a confirmed role and structure from the Dibb Report, but equally many were skeptical that anything substantial would change and remained deeply suspicious of the Labor Party's commitment to defence and the reserves in particular.

Immediately prior to the release of the Dibb Report the CMFA's May 1986 newsletter summarized the state of the ARes in the context of 'rumours of what the Dibb Report might contain and the concern which we all feel about the apparent failure of the recruiting and retention campaign' which had begun in July 1985. Victoria's low returns caused Lowen to wonder 'whether Victoria feels it is so far from the front [meaning Queensland and Western Australia] that there is no danger'.[176] The Dibb Report was released in June 1986. 'The emergence of the Defence of Australia doctrine as the foundation upon which Australian defence policy would be based following Dibb›s review of Australia›s defence capabilities, seemed to provide the Reserves with a definite role.'[177]

The Minister for Defence, Kim Beazley, outlined in Parliament some of the impacts on the Army Reserve (comments on Navy and Air Force reserves were also included) from the Dibb Report:

> Mr Dibb makes a number of major recommendations about the reserves, designed to give them more central and challenging roles in the ADF. Mr Dibb proposes that the Army reserve should take over from the regulars some of the heavier weapons-including tanks and artillery-which are maintained primarily to provide a base for expansion. This would entrust the reserves with a vital and exciting role. But that does not mean that the reserves will only have a role in remote, high-level contingencies. It is also proposed that the Army reserve should be given specific responsibility for the security of vital installations in our north. Mr Dibb has concluded that this task would be a major undertaking even in the type of low-level contingency which could emerge with little warning.
>
> Mr Dibb envisages that individual reserve units in southern States should be allocated specific towns and installations in the north which they would be deployed to defend in an emergency. Reserve units would deploy to exercise regularly in their allocated area of responsibility, giving them greater familiarity with their wartime tasks, and increasing their sense of purpose.
>
> Similar recommendations are made for the Air Force and Navy reserves, in relation to which Mr Dibb recommends a substantial expansion. In the Air Force, Mr Dibb proposes that reservists take over specific roles in aircraft maintenance and airfield defence. In the Navy he proposes that reservists should take over the operation of mine counter-measures vessels, which, as I have said already, is one of the key areas of defence in plausible low level contingencies. The Government is very concerned to strengthen

the role of the reserves in the ADF, and I believe Mr Dibb's proposals are an excellent guide to how this could be done. As a start, I agree with Mr Dibb's endorsement of the long-standing proposal that limited call-out of the reserves should be possible in situations short of a declared defence emergency. I intend to take legislation to this effect to Government. When implemented it will do much to increase the value and flexibility of the reserves as an integral part of the ADF.

These new tasks and directions for the reserves will require a greater training effort, and we will be considering increasing the number of days training each year. We will also consider an increase in Army Reserve numbers from the current figure of 23,000 to a new ceiling of 26,000 by 1988, instead of the present ceiling of 30,000.[178]

Major General Lawrence ('Laurie') G. O'Donnell AO, the GOC Field Force Command, was invited to address the CMFA NSW AGM in Sydney on 26 June 1986. Understandably, he gave a strong speech supporting the various changes underway—integration, 'affiliation' and the effects of the Dibb Report on the Army Reserve. His equally strong endorsement of the Government's public stance in support of the Reserve was also understandable given his position, but possibly received with a little scepticism by the audience.[179]

CMFA responses came in an additional newsletter in July 1986 (note the newsletter went to councillors and life members and presumably left it to branches to disseminate further); 'Looked at from the point of view of the defence of Australia, [the Dibb Report] is a rather pathetic solution, based on a far too optimistic assessment of likely threats and warning times', wrote Lowen. He did not hold back—'It appears to be a noble effort to conduct an Appreciation backwards, starting with the Plan and arriving at an Aim and Factors to justify it. I feel sorry for Paul Dibb who had to work within such constraints.'[180]

Lowen went on to analyse the report:

> From the ARes point of view, it is a mixture of good and bad. The recommendations on call-out legislation, provision of specific roles for units and return to tax-free status for reserve pay are all excellent, and overdue.
>
> On the other hand, only an academic or a politician could present a reduction in target numbers of 4000 as an increase of 3000. There are also indications that the ARes may be used as merely reinforcements for the regular units which are understrength.
>
> The recommendations on Armour, Artillery and mechanization generally are impossible to understand, however they may result in increased opportunities for the ARes in these fields.
>
> It does appear that the [Government] White Paper currently being prepared will follow most, if not all, of Dibb's recommendations. My concern now is that the Government may do what was done with the [1974] Millar Report, namely implement all the bad things but not the good.
>
> It must be our role to see that this does not happen.[181]

At the CMFA AGM in August 1986, at Victoria Barracks in Sydney, the meeting was addressed firstly in the morning CESRF/CMFA joint session by the Minister Assisting the Minister for Defence, John J. Brown MHR (Labor/Parramatta) and Major General Adrian Clunies-Ross AO, MBE, Chief of Operations, representing the CGS. In the afternoon session came CRES, Major General Cooke and Commander 3 Division, Major General Barry.

Cooke spoke at length about the current Army plans, which, he said, 'heralded a new era for the Army Reserve'. Cooke stressed that the present opportunities 'must be seized and full advantage taken of them, otherwise that last chance we have of retaining the reserve as a viable element of the Australian Army would be lost'. Major General Barry made an important point, that the CGS had put CRES on the Army Reserve Review Committee and that this 'signalled the mood of change and confirmed the concept of the Total Army', a concept he considered to be 'genuine and a magnificent opportunity' for the ARes.[182]

Branch reports came next. Brigadier R I ('Sam') Harrison MBE RFD ED for Queensland noted that it was looking at 'corporate membership' for units. NSW under Brigadier Amos reported that membership was now 55 and rising. Lieutenant Colonel Bullard said membership in Victoria was 300 of which 94 were financial. He announced he was stepping down in favour of Colonel Brian D. Clendinnen ED. For South Australia, the State delegate Brigadier I. F. ('Sam') Barr RFD ED noted it was his first meeting and he had not even had the chance to attend a State committee meeting so had little to report. In Tasmania Lieutenant Colonel Anthony ('Tony') C. Bidgood RFD ED had reported a membership of 420; 'there is great confidence in the future of Tasmanian branch'. From Western Australia, it had, according to Lieutenant Colonel Trevor J. Arbuckle RFD ED, 200 members of which 80 are financial. The NT struggled to maintain a large membership due to limited numbers and wide dispersion, plus the very high turnover of population.[183]

Meanwhile, despite approaches to Government to allow a Reserve position to be put to the DFRT through the Defence Advocate, David Michael Quick, QC, an Adelaide barrister, was still being discussed, with no decision in sight. At the AGM in August 1986, the National Council discussed how best to get its submission before the DFRT.[184] Brigadier Lowen accepted an offer by the Victoria branch, which had been very engaged on the question of Reserve pay and allowances, to develop a position paper for the National Council to submit to the DFRT. The CMFA executive also instructed the sub-committee to research and prepare submissions, seek the views of state branches, and submit them for approval to the National Executive.

The branch set up a sub-committee consisting of Colonel Gregory ('Greg') H. Garde RFD, Lieutenant Colonel Anthony ('Tony') S. Furse; and Dr Christopher N. Jessup to work on this and report to the National Council in August 1986.[185] Jessup, with his PhD in jurisprudence and experience in industrial law and advocacy was to be the Association advocate in those cases which the CMFA could put direct to the DFRT. Unfortunately, there was a 'break down of communication' in the matter because NSW branch under Brigadier Amos had also commissioned a paper on similar lines, in December 1986.[186] The executive was taken somewhat by surprise when the NSW 'Report on the Army Reserve Pay and Conditions of Service', compiled by Lieutenant Colonels George S. Staziker RFD ED, John P. Keefe RFD (a chartered accountant) and Kenneth ('Ken') J. Broadhead RFD (a Reserve Bank officer), was received in May 1987.

There should not have been any confusion, as Amos and Lowen had exchanged correspondence on the matter in September and October 1986, although Lowen seemed to think NSW branch would comment on the Victorian submission rather than complete their own submission. Lowen made an interesting observation to Amos in a letter of 10 October 1986:

> Between you and I, my personal feeling is that we should be very careful we don't allow this thing to develop to the stage where we look like a soldier's union. Naturally we are interested that the members of ARES receive a fair deal, but we know that any big increase in manpower costs will simply mean a smaller ARES in the end, and an effective Army Reserve must be our main aim.[187]

Did Lowen's comment indicate an underlying sensitivity, after all these years since the start-up of the CMFA, that it was still regarded as a sort of trade union?

CMFA Secretary Lieutenant Colonel MacLeod had to apologise to Amos when he admitted that the Victorian branch and Sub-Committee did not seem to be clear on the sub-committee's role on behalf of the National Council to collate State views rather than expect the other States to write their own complete submissions.[188] It was NSW branch's turn to be slighted by the Victorians; they may well have felt a lot of effort had been wasted. However, their conclusion certainly aligned with the CMFA's approach at the time:

> ...the whole system of ARes pay and conditions is inappropriate to the economic, social and industrial employment conditions prevailing in the 1980s. Changes to the ARes pay and conditions contemplated now must move with the times and must be capable of change.... the ARes should be aligned with the Regular Forces pay and conditions of service so that constant review is guaranteed.[189]

The Australian Cadet Corps was another subject of discussion at the 1986 AGM. Army Office was proposing to transfer responsibility for cadets from Field Force Command to Military Districts in 1987; and policy aspects would be transferred to DCRES. In another move towards tri-service alignment the Cadet Charter was to be re-written to accord with

Air Force and Navy Cadets. And Major General Cullen was not to be forgotten. He said that as Anzac Day 'was in decline, there was a splendid opportunity for the Army Reserve, and the Services generally, to make it their March'. By this he meant, the 'men and women of the active forces gradually taking over and continuing the traditions of Anzac Day'. The Association duly adopted the proposal.

In his annual report for the August 1986 AGM and the last under the CMFA umbrella, Brigadier Lowen made a number of observations. With regard to CMFA organisation and membership, he noted that all branches with the exception of the Northern Territory (NT) have full working committees—in the case of the NT, he said, Major Dunn did not wish to continue and was seeking to appoint a new president to rebuild the branch to its former strength. Lowen stated: It is true we do not seek large numbers of members, but it is essential that we maintain reasonable numbers in each territory to retain credibility....'[190]

Lowen said that the Dibb Report, 'despite its obvious weaknesses', seemed likely to set the pattern for the immediate future of the ARES and its role in the [ADF]. Lowen noted that some recommendations such as Call Out Legislation, tax relief and defined ARes roles, have been part of the CMFA agenda for some time. However there were more recommendations which, Lowen said, 'could be a cause of some concern...which tended to 'create the image of the ARES as a supplier of reinforcements, and a 'home-guard' role, rather than as a fighting force.'[191]

By this time ARes posted strengths had dropped below 23,000 despite an enhanced and costly recruiting and retention—tri-service—campaign; Lowen said that it 'has been a complete failure to date'. Only Queensland had shown any gains. Lowen listed the ways in which the Government had contributed to this failure, with the result of leaving 'members of the reserve wondering whether they are wasting their time'. Although the Minister has recently sent a letter to all members of the ARes promising a 'golden future', Lowen underscored the importance and positive role for the CMFA to play in 'trying to eliminate or lessen some of the causes of the decline.'[192]

Perhaps the most important decision made at the AGM was that the CMFA, formed by Major General Cullen more than 15 years earlier, officially agreed to change its name to the Army Reserve Association (AResA). Fittingly the motion was seconded by Cullen. It was well overdue; the CMF had been renamed the Army Reserve ten years ago. Through the following year some branches would refer to themselves as branches of the Army Reserve Association (CMFA), but for all intents and purposes the period of the CMFA was over although it took another year of transition before the new name and logo was confirmed.[193]

However, given that the formation of a new Australian Defence Force headquarters along with a new Chief of the Defence Force (Lieutenant General Bennett) had occurred in January 1984, the Association was already well behind the curve. It would take another five years before it changed again to become the Defence Reserves Association.

Brigadier Lowen made his conclusion:

> After two years as National President of the CMFA, during which time we have watched a considerable decline in the ARES, I must admit a feeling of complete frustration that we have not been able to achieve more than we have. Obviously, the future of the CMFA is closely linked with the ARES itself, and despite a few promising signs I find it very difficult to convince myself that the damage that has been done to the ARES over recent years will be overcome for many years. This situation represents a serious risk to Australia's security, and its ability to mobilise any worthwhile Force within a reasonable period. I am sure that the Government does not realise the risk that it is running.

Lowen's apparent pessimism reflected his frustration after many years of effort in a good cause seemingly without great success. Yet it was not quite his swansong. To general acclaim he was persuaded to stay on as president for another six months, even though he would be travelling overseas during that time. Brigadier Sam Harrison was elected as Vice-President and would carry the reins. Harrison would become president of the Army Reserve Association in turn at the next AGM in 1987.

Part Two:
The Army Reserve Association
1987–1992

5.
Interregnum

The renamed Army Reserve Association (AResA) was about to enter a period of intense activity and focus. The Dibb Report on Australia's Defence, commissioned by the Hawke Labor Government in 1985 and published in March 1986, led to significant changes including adoption of the Defence of Australia Policy. As a result the then CGS, General Peter Gration established an Army Reserve Review Committee (ARRC) led by Brigadier John M. Sanderson AM (chair of the CGS Advisory Committee—CGSAC) and Brigadier Barry N. Nunn RFD ED.[194]

The resulting Nunn/Sanderson Report was provided to Gration in October 1986 and its report circulated to senior commanders and staff for comment. This was then considered in November 1986 by the CGSAC; further work was then commissioned by Gration to an AARC implementation project team led by Regular, Colonel Peter A. Sibree, assisted by a senior Reserve officer, Colonel Joseph ('Joe') V. Johnson RFD ED.[195]

However, the final conclusions of the Nunn/Sanderson Report could not be released until March 1987 as the Hawke Labor Government had in the meantime commissioned a Defence White Paper arising from the Dibb Report. This, 'The Defence of Australia 1987' with its 10-15-year outlook, was finally released in Parliament in March 1987. Only then could Gration officially release the outcomes of the Nunn/Sanderson Report, even though some broad content was known to the CMFA/AResA beforehand.[196]

The AARC report noted that:

> Inevitably the ... recommendations will involve change, but the intention is to keep to a minimum, change to the existing Reserve structure. The Committee is to take full account of unit and formation traditions and sensitivities and is to consult with interested parties where change is intended.[197]

This report created a storm of concern and protest among ARes officers everywhere and especially among the senior retired officers of the AResA, including among them the still formidable Major General Cullen. The Association, while not fully active in all of its branches, was more mature as an organisation in its ability to lobby and advocate than it was at inception. However, as this next major chapter in its evolution was to show, there were limits on its capabilities to influence policies which it deemed negative towards the Reserve. There remained inter-branch rivalries and a divergence of views within the Association on how to respond to the challenges it would face in the late 1980s and early 1990s. The AResA struggled at times to present a united front on all matters.

Lieutenant General Gration had promised an article for the next AResA national newsletter on the changes afoot in March after the release of the White Paper—he called it 'The Future of the Army Reserve'. Its conclusion noted that 'The Reserve will now be called upon to play a more central role in the real and immediate defence problems'. However, the devil was in the detail, such as the need for 'fine tuning to our current Order of Battle (ORBAT)', which included some amalgamations and closures of long-standing unit depots.

'I know that this adjustment will be unpalatable to some, and I can well understand and sympathize with this', said Gration, but, 'I need, and the Army Reserve needs, the wholehearted support of outside groups, such as the [AResA]'.[198] As details began to emerge of the proposed changes, Gration's prediction of a reaction was soon confirmed. Not all was negative—the president of the NSW branch, Brigadier Amos, noted that 'while the 'loss' of two [NSW] battalions from the 'Rum Corps' was not supported by senior members of the branch, serving members seemed to generally support the proposals for amalgamation and integration'.[199]

Nonetheless, as minutes from the NSW branch AGM also reported:

> The major restructuring of the ARes consequent on the ARRC report was a matter of grave concern to both serving and retired members of the ARes. In particular the reduction in the number of infantry battalions and proposed linking of the remainder produced considerable reaction in certain geographical areas.

Brigadier Amos noted the many discussions and correspondence with senior Army figures on the 'unpalatable' parts of the report was to no avail; 'this activity did not result in any change to the restructuring being carried out'.[200] It was intensely frustrating to all concerned.

There were two areas of concern to the Association over some years. The first was the burden of administration, particularly onerous on Reserve unit commanders. The second was the impact of integration on training for Reserves, which placed an unrealistic demand on ARes members to manage the same level of training as the ARA. Both concerns were being addressed by Army reviews, namely the 'Army Wide Administrative Efficiency Campaign' (AWAEC) and the Review of Employment Specifications, Trades and Army Reserve Trade Training Requirements (RESTARTTR). Had the Association's concerns had any influence on the eventual reviews being undertaken? Perhaps, but also it was hard to quantify the impact of the Association on these issues when retention rates continued to fall.

Branches of the Association were still in a state of flux and internally the association administration seemed less than ideal. This was demonstrated by the March 1987 newsletter which asked state branches to inform the national secretary as to exactly who were the office holders on their committees and who were delegates to the AGM and their contact details. South Australia was clearly still struggling: 'Colonel Stanley has not been successful in reactivating this branch, which has been dormant for a long time. He intends to step down as President at a meeting of the committee at which he will be seeking a successor'. However, Western Australia 'continues to be very active'. Brigadier Lowen noted that as he would be absent overseas for several months, the now Deputy President, Brigadier Harrison would fill in for him.[201]

Three months later, in June 1987, Major General Cooke, at the time CRES, gave a presentation at the Imperial Service Club in Sydney to an audience which included many senior AResA members. In his speech, titled 'The Evolution of an Army', Cooke succinctly noted the problems which had beset the Reserve since 1948 when the CMF had been re-constituted after World War II. Cooke noted that the culmination of the Dibb Report and White Paper had built on the earlier Millar Report of 1974 to finally bring together a path forward for the Australian Army.

He also noted that the Army had already been progressing some of the policy changes, with integration top of the list. Call-out Legislation approved by Parliament in early June 1987, was the 'glue' to bring these things together, despite the challenges of meeting the new directions for the Reserve from its existing manpower base. Overall, it was a call to action to support the new policies, but whether he convinced his audience was another matter. Cooke later lunched with delegates at the AResA AGM in Melbourne that August, updating them on plans in place for changes to the ARes.[202],

At the AGM of the National Council held at HQ 3 Division in Melbourne in 1987, Brigadier Harrison chaired the meeting in the absence of Brigadier Lowen overseas and was subsequently elected as president with Amos from NSW as his vice-president (also president NSW branch). Along with this was the relocation once again of the Association executive, this time to Queensland.[203] Since the last AGM the Association had been busy—

teams of senior officers were working on papers to respond to the Dibb Report, the Government White Paper on Defence and the Army Reserve Review Committee reports. The NSW team led by Brigadier Dart, was already drawing conclusions and some were sure to be controversial.

In noting some of the team's thoughts Dart did make one important observation, that 'there was no formula for integration; it can work, but equally it can fail'. He added: 'The degree of success depends on the leadership and management skills of the commanding officers of the day.'[204] Dart's conclusions were, according to Colonel Brian Clendinnen of the Victorian branch, similar to what the Victorian branch had submitted to the Minister for Defence back in 1983!

The Chair, Brigadier Harrison, noted that the CGS would 'most definitely stand by the decisions that had been taken'. In light of that he questioned 'what course could the Association adopt if it ignored these facts, no matter what members felt about the loss of a number of infantry battalions?'. This led to some questioning the Association's purpose and existence. What was the point if all it did was rubber-stamped Army Office policies? The Association's case had been put, was the reply, but to no avail.

The discussion became more direct and pointed. An increase from 26,000 to a ceiling of 30,000 could support the extra battalions now being lost, said one officer. Another noted that the actual strength was only 22,000; 'it was pointless' trying to get Army Office to increase the ceiling. Brigadier Amos, as peacemaker, stated:

> The Army, including the Reserve, suffered endless reorganisations and this was bound to continue. ...if the long view were taken, the truth would come out and matters would probably come out right in the end. It was a case now of supporting the present situation but keeping in mind what would be a better solution and working towards that.[205]

But even Amos was still extremely concerned that the most serious factor in the reorganisation was the restriction of opportunities to promote reserve officers to become leaders of the Army Reserve in the future. This point hit a nerve—the loss of formation and unit commander positions to regular officers meant that there might never be 'red hat' positions for ARes officers at all. Eighteen months later this would become an explosion of concern, as will be seen.

Major General Cooke, as CRES, was appraised of the 'strong views,' being expressed over the reorganisation and asked for his own—he responded by saying that the Association needed to understand the structural problems and manpower problems Army faced. He answered questions. The discussion then resumed. Brigadier Harrison summed up from his perspective. He 'wanted one significant fact very clearly recognised', he said:

> There now exists a very close rapport between the CDFS, CGS, senior officers at those levels and the Minister for Defence; closer than ever seen before. We cannot use the 'back-door' approach because, unlike previous ministers, the present one has no 'old boy' net. Although we can be assured that we can put an Association view forward, we will never be able to break that nexus. ...the Council appears to want to convince the politicians against the advice of the military.[206]

The frustration emerging throughout the discussion was palpable. Lieutenant Colonel Bullard from Victoria, like his colleague Colonel Clendinnen, strongly felt that 'while the people believed we had a regular division and a credible reserve in fact there was only about a regular brigade and some reserves. If the Association was happy to accept that, it should hand [it] over to another body that will do as it is told and disband'.[207] The Council resolved to push on with its team work on the papers in train to further develop the Association's position. It is tempting to speculate what Major General Cullen, who was unable to attend the meeting, would have added to the conversation.[208]

On other issues, another NSW group headed by Lieutenant Colonels Staziker, Keefe and Broadhead addressed ARes pay and conditions, reporting in early 1987.[209] Their paper later contributed to an Association submission to the DFRT, but not before being criticised by Victoria branch that the paper 'reflected badly on the ARes, its relationship to the Department of Defence and the community'. The Association's performance on the taxation issue 'was ineffective' and 'had no positive effect on the outcome'. It was wrong of CRES to say, for example, 'that there was no point in the Association formulating a view different from the DFRT.....'[210]

The Victorian branch reactivated its sub-committee under Colonel Garde to also look at pay and conditions; it had returned to the same point it had reached a year ago when the sub-committee was first activated. Nothing had been achieved in a year. This was the sort of procrastination adding to the general sense of frustration in the lack of progress and impact of the Association in general. Garde later commented:

> When I was approached to lead a study on Army Reserve Pay and Conditions of Service, it was clear that the fundamentals of Army Reserve role, tasks, strategic framework, and organisation would be the bedrock of the study. This would permit comparability with Regular Army pay and conditions and that of other occupations. The only way to proceed was to adopt the Association's position as advised by the President. What emerged was that there was no agreed position as to key matters within the Association and no direction was ever forthcoming.[211]

Later that year the Victorian branch noticed an advertisement in the *Australian* on 28 November 1987 which called for submissions to the Joint Committee on Foreign Affairs Defence and Trade on the subjects of ARes pay and conditions and personnel wastage rates in the Australian Defence Force (ADF). They even called an extraordinary meeting attended by 40 officers to discuss the two issues.[212]

Manfred Douglas Cross MHR (Labor/Brisbane), Chairman of the Defence Sub-Committee, later admitted that little thought had been given to ARes matters; the wastage rates of the ARA were occupying the Committee's time.[213] In further developments in April 1989, Cross was given terms of inquiry into the Australian Defence Forces Reserves and public advertisements called for written submissions in addition to holding public hearings. Naturally the Association saw this as a major opportunity to put forward a range of views on the current and future state of the ARes.[214]

The AResA AGM in 1987 and the move of the National Executive to Queensland from Victoria saw Colonel Keith O'Dempsey RFD replacing Lieutenant Colonel Macleod as national secretary and Lieutenant Colonel R. J. Byrne RFD replacing Colonel Sandow as Treasurer. Major General Rodney G. Fay AO RFD ED was also appointed Assistant Chief of the General Staff—Reserve (ACRES-A).[215] NSW branch reported 65 financial members; Victoria 85. Tasmania reported 21 'corporate' members but not individual members. However, it also reported that the Tasman Exchange Scheme which it had initiated was now underway. Reservists from Tasmania would be considered 'on duty' whilst on exchange with the New Zealand Army.

Lieutenant Colonel Arbuckle from Western Australia reported 35 members and plans to produce *Western Soldier*, a quarterly newsletter to 250. There was no delegate from NT and no report. Queensland branch, in a somewhat unusual step, announced that it would not pursue individual membership but concentrate on unit membership. Seven units signed up. In mid-1988, Brigadier Harrison noted that Queensland branch had all of 16 individual members ('an increase of 10 this year').[216]

In the national newsletter of June 1988 Harrison, as the new president, noted that he was to discuss in Canberra four issues:

- The Association submission to the DFRT
- Conditions of service for ARes personnel on call-out
- Honours and awards for ARes personnel
- Necessity for annual indexation of maximum compensation for long term disability (at the time unchanged from the original $500 per week).

The issue of honours and awards was an old one for the Association. Essentially it revolved around Reserve angst that the Reserve never seemed to be given honours and awards in the expected ratio of establishment to the Regulars. Arguments and discussions around the cause of this had been going on for years. It was poor citation writing by Reserve commanding officers, said some; the fault of staff officers who were over-zealous in assessing them, said others. It was probably both of those factors at work combined with the lower profile and fewer opportunities for reservists to shine.

The 1988 AGM of the Association was held in Brisbane in August that year. Minutes from that meeting have not survived, but it appears that compared to the meeting the previous year it was relatively straight forward. However, the fall-out from the discussions in 1987 continued to reverberate. Brigadier Amos reported to his branch in NSW in November 1988 that he had asked Brigadier Harrison about the 'appropriateness' of the Association aims. Harrison, he reported, was against any 'outright confrontation' and felt that the 'correct approach was to work through the system'; the Association should not 'rock the boat'.[217] Amos, like his Victorian counterparts and many others, did not agree, strongly supporting political intervention.

Amos suggested that Manfred Cross MP might be a good person to approach to support Association goals. It was a far cry from the days when the Association president, Major General Cullen, could, and would, simply pick up the telephone and call the PM or Minister for Defence on an issue, bypassing the service chief. It appeared like the apparent failure of the Association in general to affect the reorganisation in any way, or quickly progress its papers to the right channel in a timely fashion, or even coordinate successfully the different and sometimes discordant views within Association branches and its senior members, had worn everyone out. Understandably, there appeared to be some weariness.

Arising from the 1987 AGM, Victoria branch had agreed to prepare a paper for the National Council commenting on the Nunn/Sanderson Report—highlighting the weaknesses and proposing improvements.[218] However, their effort was overshadowed in October 1988 by the AResA paper by Brigadiers Dart, Philip C. Parsonage RFD ED, and Max F. Willis RFD ED in NSW branch. The paper, 'Army Reserve Future Directions' outlined directions 5-10 years ahead and commented in depth on the ongoing re-structuring.[219] In his paper, Dart indicated that the Dibb Report was 'a flawed document', 'based on demonstrably false premises', and the Defence White paper 'fatally flawed'. Dart *et al* regarded the White Paper as a 'quite commendable document as far as it goes', but that it was a 'paper tiger' in that it made no hard financial commitment to expenditure.

The paper not only attacked the One Army concept—it flatly rejected it. 'The concepts of 'Integration' and 'one army' should be abolished in peacetime, the authors said, as these concepts 'have nothing to do with Defence preparedness and everything to do with preserving in peace the command structure of the bureaucratically orientation of the Regular Army'.

HMAS *Bayonet*, Melbourne Port Division, January 1987
(Sea Power Centre—Australia)

The report recommended an ARA of 20,000 and an Army Reserve of three divisions plus a communications zone of 60,000. A selective National Service scheme was put forward as the means to achieve these numbers. Concern was expressed at the ARes was becoming the 'handmaiden' of the Regular Army and that it would be 'cannabalised' in event of credible low level contingencies.[220] As one Victorian officer observed:

> The views of Senior Reserve Officers and members of the Association were divided. In NSW, there was a strong view that the traditional expansion base model of the Army Reserve should be reinstated. In Victoria, and other States there was more support for the Army Reserve/Total Force model. There was an element of rivalry but mostly there were strong differences about the role and future of Reserve Forces.[221]

Not surprisingly, the paper met a decidedly mixed reception. Brigadier Lowen, the AResA president, said that he 'wasn't prepared to forward the paper on' to politicians or others. Major General Cooke, the former CRES and by then a committee member in Victoria, noted that National Service was not a policy of either political party, while the Victorian president, Colonel Clendinnen, tactfully noted that 'generally State reports were not well presented'.[222] In hindsight, the views in the paper do not seem so remarkable, given the practical state of the ARes at the time and the changes underway imposed by the succession of Millar, Dibb and Nunn/Sanderson reports.

However, the way in which they were expressed *was* remarkable, especially when the paper accused unnamed senior CMF officers of being 'blinded by their own egos and ambitions'.

Strong stuff, and guaranteed to offend, unfortunately to the detriment of a cogent analysis of the flaws in Government thinking around defence matters at the time. The NSW branch newsletter of March 1989 just noted that the paper had been 'viewed favourably but [the National Council] considered it needed further editing'.[223] Thirty years later, with the publication of a history of the MUR, the frustration of the major changes in this period still resonated:

> The gradual processes of erosion and attrition were resisted by the local CMF communities, stridently and stubbornly in New South Wales and—perhaps mistakenly—more quietly and formally in Victoria, but will equal lack of tangible result...the process was akin to 'being nibbled to death by moths'.[224]

One issue exercising the Association in Victoria and in NSW for several years, and, as an outcome of the Nunn/Sanderson Report, in other jurisdictions as well, was the appointment of ARA senior officers to positions nominally at least, designated as ARes positions.[225] Policy was to place an ARA officer in command only when a suitable ARes officer was not available, but reservists took it as an affront if that happened when, in their eyes, a suitable ARes officer *was* available.

For example, when regular colonels replaced ARes brigadiers as commanders of 4 and 5 Brigade, letters started flying—to the Minister for Defence Kim Beazley, to Lieutenant General Gration and to a brisk exchange of correspondence between Major General Cullen and the CGS, Lieutenant General O'Donnell AC.[226] The Reserve was quick to be irritated, especially if the ARA was seen to be given special pathways. These, at times emotional, responses were nothing compared to the response when Major General Neville R. Smethurst, AO MBE, Commander Land Force, poured petrol on the flames in March 1989.

CRES had been holding annual conferences on Reserves matters for some time, to which the AResA president and association senior members were invariably invited. At the March 1989 conference came a critique of Reserve issues in the keynote address by Major General Smethurst. Smethurst had some incisive insights into the state of the Reserves and its problems. There was no doubt that retention issues were of major concern, especially in holding and training both senior NCOs and officers.

The Reserve was already short 350 officers in 1989/90. Smethurst had thought about these issues and brought along some considered ideas on how to correct them. However, it was Smethurst's blistering attack on the capabilities of senior Reserve officers which evoked an immediate response from the Reserve constituency. He was already on record as saying, at a CESRF conference in October 1988 that 'I will not accept ARes officers in command appointments who I would not take to war', so it should not have come as a complete surprise when he expanded on the theme.[227]

In his address Smethurst maintained that Reserve brigade and divisional commanders were 'over-promoted' and that training for higher command was unnecessary; rather the 'focus of Reserve officer training therefore needs to be adjusted downwards to train officers properly for regimental duty'. Smethurst was later widely attributed with the comment that Reserve officers need not rise above Major rank—in fact, he was clear that the Regulars were far better equipped to lead Reserve formations and large units. He also made it clear that he preferred regular officers in command and if he had to choose between an ex-Regular now in the Reserve and a career Reserve officer for a command, he would choose the ex-Regular.[228]

The Association at large, led by Major General Cullen, and serving senior officers as well, was incensed. Through Senator Michael E. Baume (Liberal/NSW) a statement was made to Parliament in support of the Reserves in which Smethurst was roasted. Baume started by asking the questions 'Are senior service officers serious about the Army Reserve and about the reserves generally? Do they fear the reserves as a threat to the permanent forces? Do they fear that a fair go to the reserves might not suit their short-term agendas for personal promotion?'. He then gave a general background statement on the major issues facing the Reserve, some of which had been brought to his attention at a meeting with the Association and CRES, Major General John D. ('Blue') Keldie AO MC, on 3 May 1989.

He then turned to Smethurst:

> Lately we have had a new attack on the opportunity of reservists to have a career in the Reserve Army, leading to formation command or above... The argument, especially as espoused, I am sorry to say, by the General Officer Commanding Land Force, Major General Neville Smethurst, goes something like this: there are too few places for brigade commanders; the reservist is not well enough qualified; we need command places for regular soldiers; readiness requirements have changed since World War II and reservists will not do.
>
> I have to say that Major General Smethurst, applying that kind of argument and that kind of policy, has ceased to appoint Reserve officers to formation command and above and has started to replace them with regular soldiers. This is the final straw. As the Army Reserve is operating today under the promotional strategy of Major General Smethurst, no senior commands are going to Reserve soldiers. Many qualified reservists have been passed over in favour of regular soldiers and, as a result, no Reserve officers serving today can realistically entertain hopes of holding high command.[229]

Major General Keldie himself was a case in point. He had left the Regular Army at colonel rank and joined the Reserve where he was promoted to major general. The fact that he was a Duntroon classmate of Major General Smethurst was not regarded as coincidental when he was promoted to command 2 Division in 1988. When he heard that Keldie was to become

2 Division commander, Major General Cullen wrote directly to the PM [Hawke] seeking intervention to stop the appointment 'to ensure that the whole morale and faith and viability of the Army Reserve officers is maintained'.[230]

In September 1988, Cullen wrote to the PM again on the same subject in a long, four page letter.[231] Cullen reminded the PM that he had seen the appointment of Keldie to command 2 Division as a 'further *serious* blow to the morale and normal promotion hopes of all Reserve Officers', and that it would be seen as 'the thin edge of the wedge ' to open up *all* Army Reserve senior promotions to ex-Regulars'. Now he complained that two Regular colonels had been appointed, at Smethurst's direction, to command 4 Brigade (Victoria) and 5 Brigade (NSW). It was a decision, he said, 'even more devastating than the Keldie promotion.' Several lieutenant-colonel unit commanders were now regulars as well; the effect on 'the whole pyramidal succession' for reserve officers is 'like an electric shock', wrote Cullen.

Cullen requested an interview with Hawke to discuss the issue. It is not known whether the interview was granted; the brigade appointments stood. Cullen even met with Smethurst before writing to Hawke in September 1988 to discuss the issue. Cullen was especially upset with the peremptory removal of Reservist, Brigadier Bryce M. Fraser RFD ED, from command of 5 Brigade six months before his appointment was to end, leading to the resignation of Fraser from the Army. Cullen, in a letter to the CDF, General Gration, in January 1989, described the interaction with Smethurst:

> When we discussed the questions of Brigadier Fraser and also the matter of 4 Brigade in Victoria, he [Smethurst] said to me, "Paul, if you continue to oppose me on this, I will destroy you". I said, "Well a lot of people have tried, including the Germans and the Japs". He said, "I don't mean it in that sense". However, he never did say exactly what he did mean, but I took it to be a rather unpleasant threat.[232]

That emotions were running high is perhaps an understatement, and although Cullen was writing for himself, he did so with the full weight of Reserve officers and to that extent, the AResA, behind him. He continued to advocate strongly, as he had always done, for the Reserve and its place in the 'Total Army'. In June Cullen had met with the Minister for Defence, Science and Personnel, David W. Simmons, MHR (Labor, Calare), to discuss the brigade appointments. He took with him Brigadier Max Willis, to show Simmons that there were suitable Reserve officers capable of brigade command. In a follow up letter to the Minister, Cullen expressed his view that the Total Army could not afford to restrict the base from which field commanders could emerge to regulars only. Cullen urged the minister to intervene before Smethurst, already 'clearly [acting] contrary to policy, 'goes too far.'[233]

In July 1989 Cullen wrote to the CGS, Lieutenant General O'Donnell, to complain again about Smethurst, who had apparently promised Cullen to replace the regular brigade commanders with reserve officers but hadn't done so.[234] He wrote again to Minister Simmons, as Simmons had not replied to his earlier letter, pushing Simmons for action, nominating a number of Reserve officers for promotion to senior Reserve positions.[235] He asked for a meeting with himself and Major General Murray, the former CRES, to discuss further. Simmons replied in August—recommendations for portion he said, 'have been made with due regard to the principle of Army Reserve preferment'.[236]

While it was later said that his actions ended Smethurst's career, which Cullen strongly denied, his advocacy and that of the Association itself certainly influenced a change of direction in the policy. Later at the 1989 AGM, formal resolutions, moved on behalf of Major General Cullen, were passed regarding the selection of ARes officers as commanders of ARes units and formations for the attention of the ministers on both sides along with the CDF and CGS.[237] The CDF and CGS both listened and changes were made, even though the CGS urged Cullen to rather 'channel his energies towards trying to win more resources for the Army'.[238]

In recognition of the need for the Association to move this issue and others to a higher level, it was suggested that a Defence Forces Policy Board be formed to speak with unity for the three Services and to 'plug in' to Defence Central; the Army Reserve Advisory Council (ARAC) was seen as 'little used'. One long-serving but junior officer at the meeting described the AResA as being seen as a 'toothless tiger' and 'not able to follow through'.[239] Was it a sign of frustration or simply a reflection of the generally slow reaction of the Association and its committees to change? ARAC, consisting of CRES and six others, was not actively working and on point of disbandment; perhaps a DFRAC was needed instead?

The Millar Report of 1974 had led to the formation of the ARAC. In fact the Victorian branch had already submitted a paper to the Cross Committee in May 1989 on the subject 'Formation of Defence Force Reserves Advisory Council', calling for the establishment of a DFRAC to 'give timely and pertinent advice, initiate thoughts and ideas that might not otherwise emerge through the normal command structure and generally assist in a positive manner the better development and utilization of the Reserve assets of the RAN, Army and RAAF'.[240] What were the expectations of the Victorian branch in submitting such a paper? Was it sanctioned by the National Executive? Was it any wonder that some officers were disillusioned?

Meanwhile an interesting side development within the NSW branch had been underway since mid-1988. Major General Sharp, who had joined the NSW branch of the Association in August 1985, put forward the suggestion that a Defence Force Reservists Business Association be formed. He tabled an outline paper. Sharp pointed out there was no 'all ranks' facility or association for present/past service reservists to provide 'mutual support in business, trades etc.'. He proposed that the new association could provide a membership directory of services and some club facilities possibly through the Royal Exchange of Sydney—'such a body', said Sharp, 'would also be of considerable assistance to the [CESRF] as well as the [AResA]'.[241] The NSW branch of the AResA resolved to lend support to the idea.

Months later, it emerged that Major General Sharp's inaugural committee discussions with Navy and Air Force, although they had 'enthusiastically embraced the project', was progressing slowly.[242] The NSW president then reported to members that the Sharp Committee had changed direction from a commitment to a business association, to seeking to encourage service in the Reserves. Brigadier Harrison, for one, was not in favour as he considered it to be an organisation now competing with the Association itself.[243] Several months later, Sharp's committee evolved into the Association of Australian Reserve Forces and set an inaugural meeting for June 1989, with 30 members. The 'aims and ambitions' of the new association were not clear.[244]

At the inaugural meeting it was to become The Reserves Association (TRA). By September 1989 the new association, named 'The Reservist Association' (TRA), boasted 60 members. By comparison, the AResA NSW branch had 72 members but 45 were unfinancial.[245] Ahead of its time, at least as far as the AResA was concerned, The Reservist Association had Navy and Air Force representatives as well on its committee. However, questions were being asked as to the purpose of the TRA and its apparent cross-purpose with the AResA itself.

Attention was starting to focus on the proposed *Exercise Kangaroo 89*, which was expected to see the validation of the integration and reorganisation of the Army and its Reserve components. At the time, it was touted as the largest peace-time exercise since WWII. The participation and performance of the Reserve would be watched very carefully, by both supporters and detractors of integration. In the meantime, the work of the Association continued. NSW branch provided short submissions to the Joint Parliamentary Committee in June 1989.[246] While noting that 'other submissions' would reach the committee from other branches and the National Council of the Association, it is not clear whether NSW branch had cleared its submission through the National Council and other branches as had been agreed at the 1987 AGM.[247] In any event, its assertion that the Reserve should be grown to 75,000 and the Regular Army reduced to 23,000 would have undermined its other well considered points.

Its further recommendation that a tri-service Reserve Forces Council of two-star officers advise the CDF, Secretary and Minister for Defence on all aspects of the Reserve would similarly have been viewed with scepticism, given that it would bypass the chain of command.[248] Subsequently Major General John R. Broadbent CBE DSO ED, a contemporary of Major General Cullen and a Life Member of the Association, put forward a new proposal for 'identifying and dealing with significant matters of philosophy' affecting the Army Reserve.

Broadbent proposed a meeting of about 20 senior retired Army Reserve people who would identify significant problems, select the three most important, discuss in turn with a seminar of retired and current senior reservists 'in positions of influence' and resolve these issues before communicating them to Army level.[249] Was this yet another reflection of frustration that the Association did not seem to be able to cut through and influence the major issues affecting the Reserve? The NSW branch diplomatically agreed that the concept was a good one for the United Services Institution to take up.

Did the Association benefit from other ideas submitted to the Parliamentary Committee? Although the AResA president was given access to all submissions sent into the committee, from the association or otherwise, it is very unlikely that he would have time to read and digest all of them. And the association simply did not have the resources of a 'secretariat' to exploit other viewpoints which could have helped the AResA come forward with some more progressive ideas for the Reserve than it did. While the various papers submitted from serving or recently retired senior Reserve officers like Fay, Cooke, Keldie and Murray as well as Cullen and association seniors were available to the Association—and generally were aligned in what they had to say—others also had some good thinking in them.

An example is Submission No 73 (of 111 sent into the committee) prepared by a senior recruitment consultant, Ian Mott, in the Sydney firm of Von Nida Dews and Sachs Pty Ltd. While no doubt having an eye to possible employment as a recruiting consultant, the firm had clearly considered the Reserve problem of recruitment and retention very carefully especially in light of its new 'northern' responsibilities as laid down in the White paper and Dibb Report.[250] As we have seen the Association's determined and essentially highly conservative focus on growing the Reserve at all costs (usually at the expense of the Regulars) was probably counter-productive in the face of the changing political environment.

Mott presented his views directly to the committee at a public hearing in Sydney in February 1990, the same hearing as Reserve Major Generals Cullen, Fay, Keldie, Cooke, Murray and Brigadier Amos. In the consequent national newsletter, Amos described Mott's views on recruiting and retention as 'some controversial notions'.[251] In September 1989

Major General Fay, the then ACRES-A, addressed the NSW branch at its AGM, presenting his views of the Government's White paper. His presentation was based on his submission to the Cross Committee inquiry into the Reserve which he had shared with Brigadier Amos before appearing before that committee in person.[252] He made the point that there were four important differences with the Reserve in 1989 compared to 1939:

- The size and composition of the Regular Army directly impacted on the way the Reserve was structured, trained, and developed
- Self-reliance needed a balanced force—defence could no longer rely on a larger allied force to provide logistics
- There was a much shorter time between call-out and commitment to operations
- Call Out Legislation means the Reserves will be committed.

Fay stated that 'integration is now successfully in place at all levels of the Army...' and that he believed that 'in time, the Total Force concept will ...eradicate the rivalry and jealousies of the past....'[253] Among Fay's audience however, there might well have been officers still smarting from Smethurst's attack a few months earlier, and many who remained sceptical of progress being promoted by serving reserve officers in this regard. Fay himself was an advocate of 'best officer' for the job of command regardless of origin. A year later, Fay retired and was elected president of the NSW branch of the Association.

6.
Future Directions

A few days prior to the AResA AGM in Brisbane on 7 October 1989, Major General Cullen also made a lengthy submission to the Cross Committee. Its main aspect, said Cullen, was 'the importance of nurturing the Reserve Army so that it can play its essential and required part in the Total Army'.[254] The length came from copies of a series of letters and related information he had written since November 1987 in pursuance of this theme. Cullen asserted that the regular officers, under pressure themselves from a shrinking budget and therefore smaller regular army, did not 'understand the ethos that makes the Reserve capable of playing its full part'.

Cullen also noted that 'in recent years these traditional and revered mores have been ruthlessly pushed aside, to the detriment of the morale, and therefore the efficiency, of the Army Reserve'.[255] In a response to Lieutenant General O'Donnell's letter to him of 14 September 1989, in which O'Donnell had decided that more emphasis had to be given to 'the selection, training and preparation of Army Reserve officers', Cullen wrote back on 3 October to say that 'My colleagues and I will continue to be concerned until we observe that the forthcoming initiatives are actually put into place by you [O'Donnell] and your successor.'[256] In his usual forthright manner, Cullen was not one to let the Regular hierarchy 'off the hook'.

The October 1989 AResA AGM and 'conference' (this is the first reference to an Association 'conference') had been deferred from August due to the prolonged airline pilots' strike that year. Despite the postponement, still only four States were represented; WA branch delegates came via Singapore! At this AGM there was a return to NSW for the National Executive. Brigadier Amos was elected President replacing Brigadier Harrison, Major Patrick R. Donovan AM RFD ED replaced Colonel Keith O'Dempsey RFD as Secretary and Lieutenant Colonel Robert Hill replaced Lieutenant Colonel Byrne as Treasurer. Major General Cooke was also elected vice-president.

By this time the WA branch under Major Brian W. Gibson RFD ED had 16 individual and 17-unit members. NSW branch had come roaring back from being almost defunct to 70 members, although only 43 were financial. Life membership was conferred on Major General Kenneth Eather, Brigadier Leslie D. King OBE ED and Brigadier James E. G. ('Sparrow') Martin CBE DSO ED. [257] In a footnote to earlier concerns about Reserve appointments, on 1 April 1990 Warren Glenny was promoted major general and appointed as Commander 3 Division. He was the first commander from NSW to be appointed to the Victoria-based Division, but he was at least ARes.[258]

At the meeting, the subject of entitlement to Vietnam campaign medals for ARes Observers was finally settled—they were not entitled (although a loophole saw them awarded Returned from Active Service badges). As usual the perennial issues around conditions of service were discussed yet again, although Amos made the point in a later newsletter that:

> There are many differing views about the relevance and importance of monetary conditions of service as far as the Army Reserve is concerned. A strong view is that any funds available are better devoted to equipment and supplies for training rather than pay changes. Likewise, any moves which might jeopardise the tax-free status of Army Reserve pay must be avoided.[259]

Amos gave much more importance to the Military Compensation Scheme (MCS) as COMCARE for Commonwealth employees was both 'inadequate and inappropriate' for ARes members with their 'inherently more hazardous employment' conditions than public servants.[260] The WA branch was tasked to investigate with WestPac Bank, which administered Defence Force housing loans, whether reservists could access these loans as a retention incentive. Access to retirement benefits was also to be further investigated; a discussion which had already been underway for the past six years without any outcome.

The AGM was addressed by Major General Fay who spoke on the directions being taken by the Army Reserve. It was a speech as much about the issues facing the reserve as about progress in integration and training being made. Problems loomed larger—with equipment shortages affecting retention (50 percent of soldiers were still being lost in the first two years of their service) and training, shortage of officers, a cumbersome pay and administration system, regular positions unfilled, and the list went on.[261]

Brigadier Amos reported that there had been over 280 group and individual submissions to the Joint Parliamentary Committee enquiry into the Reserves, and the Committee chairman, Manfred Cross, attending the AGM, invited further submissions. Cross said that the credit for the enquiry was in many respects due to the former Minister for Defence, Kim Beazley, who according to Cross, believed that insufficient attention was being paid to the ARes. Those at the meeting would have generally agreed with the sentiment.

However, given the outcome of that Minister's tenure, overseeing a real drop in the defence budget of around $1 billion and the further diminution of the Reserves capability and role, others looked askance at the credibility of the statement.[262] By that time, it was also reported that Beazley and the CDF General Gration had also made adverse comments about the Reserve's performance on *Exercise Kangaroo 89*, saying that it was not combat ready—another 'black mark' against the minister among those always ready to take offence.

Reflecting on the Sydney conference of 1989, Association national secretary Colonel O'Dempsey said:

> There were concerns too that the size [of the Association] was not without its costs. Possible dangers were seen in dissipating energies in hyperactivity and organizational maintenance rather than in pursuing our primary goals. The feeling was expressed that supportive COs and commanders were a better option than large numbers of 'card carrying' members. The cutting edge and leverage would always be provided by influential persons. The Association needs to speak for a large constituency, but this does not necessarily mean we must have a large number of individual members.[263]

The comment about 'dissipating energies' was an accurate one. As we shall see, the capacities of the Association were becoming almost overwhelmed as it tried to respond to the wide range of issues being considered that year and next. Some at the meeting were starting to question how best to make submissions as branch submissions were leading to contradictions with Association submissions.

In a discussion about incorporation—to give the Association a legal identity—at the 1989 AGM, the existence of the 'other' reserves association already incorporated in NSW—the TRA—was raised. This association, led by Major General Sharp and described as a 'mutual self-help body' was seen with its name and activities to be confusing the picture viz the Army Reserves Association.[264] The issue would not be resolved for another three years.

Call-Out legislation continued to be an issue of high interest to the Association. As Brigadier Amos stated in his national newsletter in December 1989, 'the implications of this legislation are important'. He noted that sensitive issues were being addressed by a working party comprising Brigadier Graeme B. Standish AM RFD ED as Chairman, Lieutenant Colonel Michael J. Finnane RFD QC and ARA member Lieutenant Colonel Martindale (fnu).[265] A questionnaire was circulated to 10% of the ARes up to lieutenant colonel and all reservists above lieutenant colonel rank, with a report to be furnished in early 1990.[266] The issue of Call Out Legislation and how it impacted reservists had been an issue for several years already and there seemed to be no end in sight to the endless reviews and delays in implementing it.

During February 1990 the Cross Committee was holding public hearings in Canberra, Sydney, Brisbane, and Townsville. In Sydney, submissions were taken from several senior ARes officers, summarised in the AResA newsletter that month. Major General Cullen stated his view that 'there were within Army Office a group of officers determined to downgrade the Army Reserve and its leadership'. Major General Fay believed that 'with more flexibility it was possible for the Army Reserve to achieve significantly enhanced capability'.

Major General Keldie spoke of the 'serious constraints to effective training imposed by current and diminishing resources'. Others such as Major General Murray noted that the current effectiveness of the ADF was jeopardised by inadequate funding; he proposed 'a re-allocation of Defence funds by reducing ARA manpower by 5,000 and replacing this with Individual Emergency Reserves and ARes to total 10,000'.[267] Brigadier Amos covered 'the generation of community support for the ARes and appointment of unit and formation commanders.'[268]

With the dissolution of Parliament in late February 1990 for the Federal election, the Joint Committee on Foreign Affairs, Defence and Trade ceased to exist as did the Defence Sub-committee chaired by Cross—and therefore the inquiry into the Reserves. However, the new Minister for Defence, Senator Robert F. Ray (Labor/Victoria) subsequently re-triggered the inquiry, and a new Defence Sub-committee was formed under Eamon ('Ted') J. Lindsay RFD MHR (Labor/Herbert).[269] Calls for new submissions in addition to the previous 115 submissions (including from the Association, NSW and Victorian branches as well as a number of AResA members) to the Cross Committee were soon called for.

The Association quickly made a supplementary submission to the new committee covering cadets, Chinooks, the Military Compensation Scheme (MCS) and 'Future Directions'.[270] In October 1990 the Victorian branch attended the public hearing in Melbourne to present its views. Meanwhile the Government was expecting yet another report at the end of March 1990 from Alan K. C. N. Wrigley AO, a former Deputy Secretary for Defence and head of the Australian Security Intelligence Organisation (ASIO). The report—'The Defence Force and the Community: A Partnership in Australia's Defence', would not actually be tabled until October 1990 after an Inter-Departmental Committee had reviewed it.

In a symbolic but important step forward for the Association in March 1990 the AResA president was invited for the first time to attend the annual ACRES-A conference in Canberra (colonels and above).[271] The same month the new AResA president, Brigadier Amos, sent out the 'Future Directions' paper to all branches and life members, with a questionnaire to all branches for comments. The 'Future Directions' paper covered a wide range of topics. The questionnaire covered four areas of interest—conditions of service, public relations, training, and equipment; each section contained between five and 18 questions. By end of April deadline however, only four branches and seven life members had responded. The NSW branch response finally came in June.

It wasn't until September 1990 that the NSW branch response was circulated to its members; in 1990 computers were still relatively new, not to mention email.[272] Essentially Amos was trying to get the Association membership, and through them their committees, to focus on what was important to make real progress. He urged the Association to:

1. never address a problem which is not seen as such by the active Reserve

2. concentrate more on problems at government level rather than in-service affairs, e.g. the Veteran Entitlement Act/Military Compensation Scheme cover, receipt of 3% productivity and DFRDB cover etc.

3. improve communications between the Association and units and boost membership, e.g. by establishing a volunteer contact in every unit[273]

The Association felt that it suffered from a lack of active support from serving reservists. In a series of 'contact meetings' particularly with staff of 2 Division starting in August 1990, the Association met with 'an aim of the meeting is to convince the Staff about & what the Association can achieve for the Army Reserve.'[274] That this was required at all was an indication of the relative failure of the Association to demonstrate its relevance to many serving officers.

The August 1990 AGM, held at HQ 2 Division in Sydney, spent considerable time analysing the responses to the 'Future Directions' questionnaire although given the rather poor response rate and the standard of those responses it must have struck the executive that they were in some respects 'chasing their own tails'. By this time, in addition to the review underway by the reconstituted Joint Parliamentary Committee and the release of the Wrigley report, yet another report on career/personnel initiatives to assist in staffing the ADF was commissioned by the CDF. Brigadier (Dr.) Nicholas ('Nick') A. Jans OAM and Judy Frazer-Jans subsequently released their report 'Facing Up to the Future'. The Association then developed a paper in response to that report as well as to the Wrigley Report.[275]

The Association's paper made it clear that it accepted the concept of total involvement by the Australian community in the defence of the country, and 'believes that the Wrigley report provides a framework within which such a concept could be developed'. Wrigley's document was 'thought provoking and of first rank significance', stated the Association, and Wrigley's recommendations were, 'in the main, timely and appropriate'. The Association also went on to say that 'there appears to be little doubt that a 'San Andreas' [Fault] type impact is needed to give new direction and energy to the ponderous glacier that many see as our Defence Department.'[276]

The closing remarks of the Association's critique were telling, reflecting the concern held by many in the AResA that the Wrigley Report would be in fact be buried, and would not, as the Minister for Defence Beazley put it, '...dominate the defence debate for the next two or three years'[277]:

> The author [Wrigley] who himself is a veteran of the infighting that sadly is a hallmark of the national; Defence community, has expressed his fears that his Review will be fought 'tooth and nail' by groups more concerned with their narrowly vested interests than with the defence of Australia. There is evidence already of the existence of a 'black hole' which will swallow his less palatable recommendations. That this should happen...would rank as a national disaster.[278]

The Victorian branch also made its own submission to the Defence Sub-Committee (the 'Lindsay Committee') on the Wrigley Report in October 1990, appealing to the Sub-Committee to use its 'considerable influence' to maintain the status quo in terms of Reserve unit structure in Victoria 'until such time as Wrigley had been digested and thought through'. At the same time, it stated: 'We believe it would be tragic if closed minds and vested interests are allowed to sweep away and condemn for their own reasons such a broad based review of the defence force....'[279] The 'closed minds and vested interests' were not elaborated.

Further on the Wrigley Report, Major General Fay, elected president of the NSW branch in September 1990, wrote to Amos a month later 1990:

> From a number of the public comments to date it is obvious that the Report is meeting some criticism and, unfortunately, some of that criticism has focused on peripheral issues rather than the central theme—greater involvement of the total community in Defence.[280]

Alan Wrigley spoke to meetings of various AResA branches through late 1990. '[The] interest being shown may yet enable it to avoid being buried by the bureaucracy', wrote Amos.[281] References to 'the bureaucracy' and 'vested interests' were mostly directed at the Regular Army, which was unhappy with the report, not least because it felt it had not been consulted by Wrigley. One commentator later wrote: 'the main theme—reduce the permanent strength of the ADF and raise a militia force—will no doubt increase feelings of unease in the Defence community'.[282] As can be seen, the Wrigley Report in some respects polarised Regular vs Reserve perspectives even further. Inevitably the tension between financial constraints and the strategic outcomes of these various reviews and reports could not be neatly resolved for the Reserves let alone the Army or ADF, leaving the AResA much to ponder on.

It was a busy time for everyone with some branches stretched to the full to meet the objectives of the Association. With so much going on demanding its attention and effort to assess, consider and respond to, it was not surprising that the Association had probably reached the limit of its capacity to do so.[283] In an attempt to manage the workload and demands, all branches were allocated tasks at the August 1990 AGM meeting, with work being undertaken across all areas of concern to the Association.[284] A snapshot of issues and responsibilities shows the following:

1. National—Response to the Jans paper and the Wrigley paper by a sub-committee with comment on draft by branches.
2. Queensland—Association internal PR to increase membership and funding; material preparation for media in support of ADF; other plans to increase public awareness of ADF reserves; distinctions between ARA and Reserves and how these should be addressed in organisation and training
3. NSW—Compensation arrangements; clothing and personal equipment and unit equipment issue
4. Victoria—Bonus schemes for efficiency and retention purposes; association policy statement on proper career paths for ARes officers and NCOs
5. Tasmania—Revival of cadets
6. South Australia—adequate provision of rifle ranges with ready access for ARes training
7. Western Australia—implications of Call Out Legislation on employers and recommended legislation provisions to cope with these.[285]

NSW branch reported:

> You are no doubt aware of the ups and downs of the AResA (CMFA) over the decades. NSW almost disappeared a few years ago but we're pulling up now...We believe the AResA will be of growing importance to lobby politicians and where appropriate go to the senior defence people. Further, as the WWII ex-service community fades away, the AResA must be set to take a more vigorous place in that scene...[286]

The association agenda and work schedule were certainly ambitious but was also to test the Association and its resources like never before.

In late 1990 the Reserve was once again seemingly under heavy criticism, this time from the Auditor-Generals' (A-G)'s Report on the Army Reserve, published in August 1990. The report generated by the Australian National Audit Office (ANAO) for the A-G was not complementary to the Reserve, saying it was expensive and neither efficient nor effective. A supporter of the Reserve in Parliament, Senator David J. MacGibbon (Liberal/Queensland) however, responded to the A-G report, pointing out the faulty methodology employed by the ANAO to reach its conclusions. MacGibbon called for the establishment of an Inspector-General of the ADF. MacGibbon went on to question a whole range of ANAO conclusions and the methodology used to reach them. [The ANAO] said MacGibbon, 'again…misses the point that the Reserve is not an end in itself but a means to an end'.[287]

MacGibbon also pointed out in his speech that the timing and conclusions of the A-G report were curious. Conducted between March and December 1989, the report was published in August 1990 only after extensive commentary and recommendations by Defence along with wide circulation within Army before publication. This, having the effect of being 'used to undermine the Reserve within the Defence community'. MacGibbon stated that 'It is not only the Regular Army but also some Regulars in the other two services who are now saying that the reserves are a waste of money'.

Even more curious was the timing of the release of the report, which was released just two weeks before the decision to disband 3 Division HQ was made public. 'Rightly or wrongly', said MacGibbon, 'there are those in the Reserves who believe they were treated less than honourably by their colleagues in arms, their brother officers, in the Regular Army'.[288]

The decision to disband 3 Division HQ and its Intelligence unit as well as merging of two artillery regiments and two light horse/mounted infantry units in Victoria was announced. The Association railed against this decision, with letters of protest and faxes sent to the Minister for Defence Senator Ray, the Minister for Defence Science and Personnel, Gordon Bilney MP (Labor/Kingston), Shadow Ministers Senators Peter Durack (Liberal/Western Australia) and Jocelyn M. Newman (Liberal/Tasmania), and to the Chair of the Parliamentary Defence Sub-Committee, Peter J. Lindsay MHR (Liberal/Herbert).[289] Ray announced a saving of $9.3 million but the Association asked how that could be the case when the minister also announced that the 50 ARes and 50 Regular staff involved would be re-deployed elsewhere?[290]

Association president Philip Amos also arranged for all living past and present major generals in NSW to dine with the CGS that month—in attendance were Major Generals Broadbent, Cullen, Fay, Keldie, McDonald, Maitland, Murchison, Murray and Sharp. While the disbandment of 3 Division was a major matter of discussion, and a public statement of concern signed by the eight retired generals was sent to the PM, various ministers, CDF, CGS and media, the Government would not review the decision.[291] In a letter to the Prime Minister in January 1991 from Major General Vickery, a Life Member of the Association, asking for the PM's personal intervention to halt the 3 Division decision, he said 'This is a dog's breakfast'.[292]

During the period that Brigadier Amos was national president, he and the national secretary and treasurer would often attend NSW branch meetings and the NSW branch committee at times acted like the national committee for Amos. This had been a major factor in the past which led to the demise of the NSW branch. However, as the Victorian branch had discovered when Brigadier Lowen was the national president based in Victoria, and after criticizing the NSW branch of failing to separate the two, it was very hard to do so when the executive had no real resources.

Consequently, it became hard to discern when the NSW branch was discussing State or national matters at their regular committee meetings. Inevitably other branches started to feel that they were 'out of the loop'. Nonetheless the national newsletters attempted to keep everyone informed, even if, 'unfortunately most of [the news was] bad' from the ARes point of view.[293]

No. 25 Squadron [Commanding Officer Wing Commander G J Ennis DFC] in 'working dress' on exercises at RAAF Base Learmonth in 1991—the Squadron's traditional Black Swan tail emblem is clearly displayed on its Macchi aircraft.[294]

(Susan Anonstrom)

In February 1991, Major General Cullen gave the Blamey Oration at the Royal United Services Institute (RUSI) in Sydney, which he called 'One Army'. Although not speaking on behalf of the Association directly, Cullen's themes touched on the many issues in which the association was engaged with over the past 20 years since he had founded the original CMF Association. A statement at the start of his presentation—'The fact is that the One Army is in a state of flux—almost of disintegration'—set the tone for what he had to say. In a *tour de force* of the history, current situation and future of the Reserve, Cullen was at his best. As always, he was passionate about the Reserve but even more so for One Army.

Cullen wondered aloud what form the Army Reserve would take—'Will they be hewers of wood and drawers of water for the Regular Army and for round out?', he asked, 'Or will they be a cogent, efficient force component, as our Australian ethos and history indicate?'. Providing 'many examples of the denigration and destruction of the Army Reserve', especially since 1987, Cullen noted the result—'an ever-increasing diminution of its capacity and its prospect for growth and increased status'. Cullen concluded that the 'unfortunate reduction in the size of the Regular Army and its number of senior officers has caused much if not most of the abrasions which have wounded the Reserve'.[295]

A change to the Association name was mooted in a letter from Amos to all members in February 1991, following resolutions from several branches led by NSW to move with the times and to reflect Defence Force structures in place. The former CMF member of the Military Board had been replaced by the Assistant Chief of the General Staff—Reserves, Army (ACResA), who reported to the CGS. The CGS in turn reported to the CDF who reported to the Minister. And with the increasing importance of HQ ADF, all operational questions devolved directly to Land, Maritime or Air Command and the CGS, CNS and CAS became responsible mainly for raising, training, and supplying units, but not for their deployment.

Amos noted that 'Any action to support Reserve interests had to be outside the system and affected at a political level—there was no direct line of access within the system for the reserve point of view to be put to the Minister'. To this end Amos proposed that 'because Naval and Air Force Reserves, and because all ministerial considerations seem to be tri-service these days', the Association take in Naval and Air Force members. Perhaps in recognition of the relative failure of the Association to fundamentally affect any major decision regarding Reserves, and in order to be an effective lobby group, it recognised that it needed stronger tri-service membership (and more funds) to demonstrate that it has *gravitas* with an expanded support base in a tri-service defence environment.

Accordingly, it was proposed to change the name of the Association to the somewhat cumbersome 'Association of Defence Force Reserves of Australia'.[296] Amos wanted answers back on the proposal by April. By June it appeared that there was a 2:1 move in favour of the name change but also considerable opposition. The proposal was postponed for further discussion until the AGM set for Randwick Barracks Sydney in August 1991.[297] The agenda for the AGM was given under the title of 'National Conference 1991'.

At that meeting there was agreement reached that the Association would henceforth take in Reserve members from Navy and Air Force, although another interim name—Reserve Forces Association—was put forward. Subsequently it was agreed that the name should be the Defence Reserves Association (DRA). Major General Fay, previously the NSW branch president, succeeded Brigadier Amos as the national president. Branch reports at the AGM noted that Queensland had 20 individual and 17 corporate or unit members (it had tried and failed to establish a sub-branch in Townsville); Victoria had 45 attend its last AGM, while Western Australia had 70 members. The NSW branch had by then regained its former strength reporting more than 130 members and 24 life members in its ranks. The name change and logo change would be formalized at the August 1992 AGM and meeting; *Citizen Soldier* would become *The Australian Reservist*, but the first edition would not make its debut for another six years.[298]

The South Australia branch reported to the AGM that it was in effect, non-existent. There was a 'steering committee' which had been meeting monthly since May 1991, 'charged with trying to generate interest in the Association'. Prior to that two meetings were held to discuss the formation of a branch. The first, in July 1990, had attracted 12 Reserve members past and present, while the latest in September 1990 had attracted only seven members. Lieutenant Colonel Murray E. Alexander RFD ED, who had chaired these various meetings and the steering committee, asked for help from the other divisions, noting that 'the response to the idea of a branch is very, very lukewarm.'[299]

The Minister for Defence Senator Ray finally made his long-awaited statement of Government policy which covered the Force Structure Review on 30 May 1991. The CGS noted that he would be providing more details regarding the review, the Ready Reserve concept and commercialization and civilianisation of Defence. The ministerial update noted that 21 infantry battalions would form the core of the Army's combat capability—four Regular, three Ready Reserve and 14 Reserve. The Association had written to the Prime Minister before the decisions were considered by Cabinet, suggesting that they were premature given that the Joint Parliamentary Committee enquiry into the Reserves had not yet reported, but to no avail.

After the Minister's statement the Association made a widely distributed media statement which also went to the four ministers on both sides—without any response whatsoever, politicians and media all. Perhaps it was not surprising given that Amos noted that the Government had 'not taken the opportunity to provide the reserves with regular support, funding, equipment and resources to enhance their effectiveness, [nor] introduced incentives to assist retention and recruitment.'[300] Amos also arranged for a dinner with Senator Durack along with Major Generals Cullen and Fay to discuss Association concerns; only platitudes resulted.

The Association made a submission to the Parliamentary Defence Sub-Committee regarding the Minister's statement on the Force Structure Review on 18 June 1991. Its opening and closing statements are illustrative of the Association's position at the time:

> This Association is concerned that Government funding of Defence continues to fall significantly short of that required to meet the requirements of DOA 87 [Defence of Australia 87—the Government White Paper]. The reduction in combat numbers of the regular forces and the substitution of part-time regulars (the 'ready reserve') does not allow the Defence Force to meet the Government's stated requirements and is at best palliative.

It concluded:

> It is disappointing that once again the regular defence forces have failed to take the opportunity to enhance the effectiveness of the existing defence reserves, by providing them with regular support, funding, equipment, and resources. The provision of such resources to...the 'ready reserve' demonstrates that there is no will or desire by the regular component to improve the existing reserve—the citizen forces of this country, which have served the country so well in two world wars. Such an attitude is to be deplored.[301]

The Victorian branch also made a submission to the Parliamentary inquiry into the Reserves through the National Council, a few days later. It was highly critical of the proposal to form a Ready Reserve and disband 3 Division HQ among other issues and questioned whether any changes would become effective without the commensurate increase in funding and resources to make it happen.[302] Like many other submissions in the past, such as the Association's supplementary submission on the Wrigley Report to the Defence Sub-Committee in November 1990, these submissions were acknowledged but little more.

NSW branch minutes from the 1991 AGM in September reflected the developments underway both within the Association and in the Reserves generally. Approaches were to be made to Navy and Air Force reservists to help bring the agreed name change to reality; the CESRF was to be renamed the 'Reserves Support Committee' (although it was hoped that 'Defence' could be added as the prefix); and the very name Army Reserve was changing to 'General Reserve' and 'Ready Reserve', (neither with the 'Army' prefix).[303]

On 29 October 1991 at the Melbourne Town Hall, a momentous event occurred in the history of the Reserves—a dinner to mark both the 75th Anniversary of the 3rd Division, and its disbandment. The Association's deputy president, Major General Cooke, made the address at that dinner. The speech acknowledged the 'hurt and sadness' that the disbandment engendered in reservists especially in Victoria, the strong 'sense of disappointment' in the decision which appeared to have been rushed through, seemingly to pre-empt the findings of the Joint Parliamentary Committee enquiry into the Reserve. It was a low point at a time when so many reservists remained deeply concerned about the overall defence capacity of Australia and its capability to expand in times of war.[304]

With publication of the report of the Joint Parliamentary Committee into the ADF Reserves the Association found areas which it could support and others it could not in the report. However, as Major General Fay reported in his February 1992 newsletter, 'Importantly it has focused on the problem of lack of resources and in this regard has set the record straight that this has been the single most important issue affecting the capability of Reserve units and formations.'[305] Peter Lindsay, the Chair of the Committee, said in his summary of the report:

> The Committee has no doubt that the military capability being generated currently by the Army Reserve is grossly inefficient. In Army, Ready Reserves will be introduced at the expense of one of Army's only three regular brigades—and essentially without any consideration of 26,000 General Reserves. The Force Structure Review has reduced these reservists to 'third XI status.'[306]

0Reports by the branches of the Association reflected this summary. The NSW branch of the Association made a submission to that inquiry as did the Victorian branch. Victoria noted in its annual report for 1991 that in its submission to the Joint Parliamentary Committee, whose terms of reference had been extended to include the Ready Reserve:

> ...our broad thrust was to decry the proposed raising of a 'third army', when it was apparent that defence cannot maintain a '3rd Division'. ...introduction of a third element would destroy the work done previously by many to remove the 'two army' concept in people's minds, but would instead promote a 'three army' view......resources proposed for the Ready reserve would be better spent if allocated to raising the effectiveness of the existing Reserve.

ARes Brigadier Standish also prepared a report for Army Headquarters following an examination of the problems of preserving the employment, social and financial status of ARes members in the event of being called out which was submitted to Army in June 1992.[307] A feature of the report regarded a tribunal system to resolve financial problems, e.g. pay differences, moratorium on mortgages etc.

RAAF reservists of No. 21 Squadron on parade, 1991
(Air Force Image Archive Collection)

Colonel Sandow led a DRA letter writing campaign—'a 'ginger' group which drew the ire of the military establishment'—to criticise the Labor Government for neglect of Defence and Reserve in particular.[308] Most of his letters were not published but they remain in the archives and give some flavour of the depth of feeling at the time, and the level of frustration that nothing was changing:

> To the *Herald-Sun*—'Now that the Parliamentary Committee report about the Army Reserve has at last been tabled [29 November 1991], showing the disastrous result of many years of neglect, it must be obvious to all that the load of rubbish about defence preparedness can no longer be tolerated.'[309]

> To the *Courier-Mail*—'There has been a concerted effort on the part of the s (the Army in particular) to have us believe that 'they are ready for any challenge'. The truth is that they are ready for very little indeed......the proposition that they are poised and ready...is a downright untruth... We have an Army Reserve system on its last legs, purely because it has been starved of equipment, stores, training time and assistance by a Government and a Defence hierarchy who are selling us all a line of rubbish....'[310]

> To the *Weekend Australian*—with a bitter attack on [Labor] Minister for Defence and Personnel Gordon Bilney for neglect of defence as 'a prominent member of the conspiracy currently dismembering our already ineffective defence forces.'[311]

In the background, another issue of growing interest was the cause of the cadets. The decision in 1983 to create Regional Cadet Units (RCUs) operating from Army Reserve depots and Limited Support Units based on school units which received no Army support, badly affected cadet numbers and therefore an important source of recruiting for Army and Army Reserve.[312] The decision, made by Army Office to save money and reduce ARA cadre staff, was, as the Reserve had already experienced, only one of a number of poorly considered reorganisations since 1975 with few positive effects. As early as September 1983 the Victorian branch had raised the issue of support for the cadets with the national president; in April 1984, the 'shambles' of the cadet training scheme was raised again in the branch meeting.[313]

It was not until Brigadier Lowen became president of the Association that the issue of cadets was taken up with any energy. Lowen asked the Victorian Committee, 'what, if anything, should be done [about cadets]?'.[314] By February 1985, the Victorian Committee (the *de facto* secretariat for the National Executive), agreed that as cadets were to be thrust on the Reserve anyway and as they would demand resources, then it needed to pay attention to it.[315] Cadets were formally placed under ACRES-A in 1987 and 'the burden of supporting RCUs was fully transferred to the Reserve.'[316]

However, there were too many other more pressing matters to deal with and while the topic of cadets continued to come up from time to time it was not the most important issue facing the association. It was not until the March 1987 AResA national newsletter that more detail on the revival of a modified cadet system was provided to members, along with details of the Association's concerns raised with the Minister for Defence. In contrast to many who wanted a return to the 'regimental cadet' system, which over many years had provided a steady stream of Defence recruits, the Minister noted:

> I would point out that the aim of the Australian Cadet Corps is no longer related to the preparation of young people for military service as was the Regimental Cadet Scheme. The aims of the current Cadet Scheme are directed toward the development of our youth including initiative, leadership, integrity and loyalty.[317]

A 1990 submission to the new Lindsay Defence Sub-Committee addressed cadets, and the Tasmanian branch took on the subject as one of its projects that same year. In the new era of the Defence Reserves Association, the cadet issue would continue to gain in importance as the Army Reserve attempted to adapt to the new direction as set out in the Minister's statement. In fact, the change towards more 'outward bound' or adventure type activities and less military type activities for cadets had been underway even before 1975. The withdrawal of Army support following the Millar Report in 1974, to save money, and disregarding the recruitment value provided by cadets was further accelerated by the 1983 decisions. The cadet system was in demise.[318]

Meanwhile, the transition to the new identity as the Defence Reserves Association was well underway. Within NSW branch it was decided not to merge with the TRA of Major General Sharp, but by April 1992 it had been agreed that an amalgamation of sorts would occur by having AResA NSW members having a majority of positions on the TRA Board—four of seven positions.[319] The first AGM of the DRA—Victoria was held in July 1992, followed by the National AGM in August that year.

The decision to move to 'tri-service' basis was by no means a unanimous decision. After all, by comparison the Naval and Air Force reserve components of the ADF were both smaller than Army's and structured differently.[320] However, with the strong endorsement of Major General Cullen, the founding president, there was recognition that the association had to adapt and move with the times, despite the reluctance shown by some of the membership. This was reflected in the speaker list for the August 1992 AGM and conference, which included presentations by Senator MacGibbon but more important from four senior Reserve officers, including from Navy and Air Force.[321]

HMAS *Adroit*, RANR training vessel, Fremantle Training Division, 1983-1992
(Sea Power Centre—Australia)

An example of the impact of changing from AResA to DRA was reflected in the report to the (now) DRA's August 1992 AGM and conference by the Queensland branch. The branch noted that its vice-president was Commander Robert W. G. Hume RFD RANR and contained a (unspecified) number of Navy reservists among its 20 individual members, but no Air Force. The 17 corporate or unit memberships only contained ARes units. The branch had tried to engage reservists in North Queensland and hoped to establish a sub-branch in Townsville but had been unsuccessful so far. 'A wider, balanced involvement of all three services in the state Association' was seen as a priority issue for the branch, including increased and significant female representation.[322] The fact that Queensland noted exactly the same number of members in 1992 as it had in 1991 was not a hopeful sign that the aspiration would be met in the coming year. Other branches would also find it challenging to attract non-Army reservists to the Association. ,

Reporting to the August 1992 AGM, Major General Fay reported that in an attempt to further improve the management of the Association's affairs, determine priorities and implement policies that a small executive group of the president, deputy president, past president (by invitation), secretary and treasurer would do this. It seems surprising that this needed to be enunciated at all, but the change to a tri-service association put even more responsibilities on the executive at a time of increasing complexity in the challenges faced by the Reserve.

At the same time, the Association leadership recognised that its limited resources and capacity meant it needed to carefully delineate the issues it could affect, and these were defined as:

- Conditions of service
- Compensation for reservists (on injury)
- Protective legislation on Call-Out (Standish Report) and
- Cadets

Fay also discussed the impact of the name change and commented: 'I anticipate that recruiting new members will not be spontaneous and will take time and effort to develop at State branch level'.[323]

Almost 25 years had passed since the CMF Association was founded. Its most tangible claims to success were the re-introduction of service medals for Reserve service, and the influence on the decision to have the tax-free pay system restored. Yet the Association, for all of that, had grown in stature over the years as it came to rely less on the individual presence of its founder and more on its own ability to work up serious advocacy submissions of quality.

Despite the challenges of establishing and sustaining a national organisation and the growing pains exhibited in the rivalry and at times, cross purposes of the two largest and most active branches in NSW and Victoria, the Association was no longer considered a dilettante in the ex-service organisation sector. It had become, through the professional interests of its members and through their dedicated and hardworking committees, more than the sum of its parts. It had become recognised by Government, and the Regular Army, as an association which demanded attention and could not at least, be ignored.

On the other hand, Fay recognised its limitations, noting in his August 1992 report that 'it was 'difficult...to identify specific areas of success...although we do provide a structure through which any matters which impact on our Reserve Forces can be voiced'. He continued: 'The continuing development of the 'voice' in terms of its credibility and influence must be one of our most important objectives...and [how] we can make our greatest contribution to our Reserve Forces in particular and the ADF in general'. Fay went on to say how the Kindred Organisations were important to that contribution, in effect admitting that the DRA would be more likely to affect Government policy through the 'considerable influence' of such organisations as the RSL and the AVDSC.[324]

Part Three:
The Defence Reserves Association
1993-2020

7.
At the Crossroads–Again

In early 1993 the DRA adopted a policy to work more closely wherever appropriate with kindred organisations to gain more support for some of its priority issues. It sought to join the committee of the Kindred Organisations Committee (KOC) of the National RSL. This body brought together some forty organisations with a Defence background although heavily directed towards the welfare issues of the Defence Forces only. Major General Fay as DRA national president represented the DRA at the KOC meeting in Canberra in March 1993 and put forward the DRA's views on Reserve compensation and the protection of Reserves legislation.[325] Joining the KOC under the RSL's umbrella indicated an understanding that the RSL enjoyed a 'louder and more influential' voice in Canberra than the DRA.[326]

During 1993, the NSW branch made a first step towards the new tri-service posture. When Major General Keldie stood down as State president and was succeeded by Major General Sharp, Squadron Leader Colin ('Col') Gilbertson, who had been involved with Sharp on the now amalgamated Reserves Association, became the new vice-president of the DRA NSW. Gilbertson had been an Air Force Reservist since 1951 and had been involved directly or indirectly with the now DRA for several years.[327] He would later become State president himself. Under Sharp, the NSW branch continued to pursue DRA objectives with renewed vigour.

For example, during 1993 the NSW branch along with the DRA and other State branches, made submissions to the Committee of Inquiry into Defence Awards.[328] Essentially, the DRA requested the committee consider Reserves for just and equitable recognition given the total number of Defence Force reservists on a serving member per capita basis. This issue was one which had been with the Association for more than 20 years. While the Association had influenced the agreement to recognise reservists for long service with their own award, the issue continued because of under-representation of reservists in awards each and every year.

Writing to the ACDF-Reserves, Major General Barry Nunn in July 1993, the DRA president Major General Fay said: 'The number of honours going to reservists is disproportionate on a per capita basis giving rise to the view that the Reserves are subject to some form of discrimination.'[329] The chairman of the review was Lieutenant General Gration, the former CDFS, an appointment which was met with dismay by several ex-servicemen. At least one ex-service organisation wrote to the responsible Minister to say that 'his appointment would be detrimental to making a fair assessment of the Honours and Awards system'.[330] No apparent changes to the distribution of awards resulted.

A new issue loomed—in July 1993 the CGS, Lieutenant General John C. Grey AO (later AC), recommended to the Minister for Defence, Senator Ray, a major reorganisation of Reserve training. He proposed moving Reserve training to Regular establishments along with the closure of many Reserve depots, a significant reduction in senior positions among Reserve officers and a diminution of the traditional roles of university regiments in commissioning new officers. Grey's plan identified 45 of 230 Reserve depots for closure, and about 15% of senior officer positions.[331] The recommendations only confirmed rumours that had been circulating of cuts to the Reserve particularly of the Training Groups.[332] This became another major concern for the DRA.

Meanwhile, the Tasmanian branch took an initiative to organise the attachment of Reserve Junior NCOs to the New Zealand Regular and Territorial Forces for a two week exercise, providing a training opportunity for reservists from Tasmania that would not otherwise have been possible.[333] The scheme was launched in Tasmania in September 1993 by ACRES-A, Major General Denis R. Luttrell AO RFD ED, and supported by local businesses. In 1994 three Army reservists one Navy reservist participated. In 1995, six ARes and in 1996 five ARes and one RANR joined in; along with, in 1995, two reservists from South Australia.[334]

It was an example of a small branch (not in member numbers but in the Reserve units it represented in Tasmania) focusing on an achievable local outcome with good results, including community recognition and publicity. Other branches struggled to get publicity to illuminate issues other than through provocative language. Colonel Sandow in Victoria observed that he was:

...happy to say that the careful placement of comment and persuasion ...enabled us to influence the decision makers on a number of occasions this year. It has regrettably been necessary to occasionally write to the press in less conservative language than we would have liked but 'needs must when the Devil drives'; when there is an urgent requirement to make a public protest and when only provocative words will incite an editor to publish a letter.[335]

The October 1993 DRA AGM and conference was attended by all State branches except the NT. Discussion at the AGM focused on several issues, including:

- The Military Compensation Scheme (MCS), designed to provide compensation for members of the ADF in case of injury, incapacitation, or death while on duty was introduced into Parliament as an amendment to the *Safety Rehabilitation & Compensation Act* (SRCA). The DRA supported amendments to the *Veterans Entitlements Act* (VEA) on the basis that all conditions of compensation should have no difference between ARA and Reserves.
- Protective Legislation on Call Out, which was subject of a review by Brigadier Standish and committee soon after the Government enacted the call out provisions in 1989. Four years later nothing has been finalised.
- Force Structure Review. Another FSR was underway, and the Minister for Defence is compiling a new White Paper due to be released by the end of 1994.[336]

Major General Cullen submitted to the AGM a policy proposal under the title 'The Reserve Forces and their Future', but it did not seem to evoke special discussion.[337] Major General Fay was re-elected president of the Association, with Colonel Don Sandow as vice-president, Major P. Donovan as secretary and Lieutenant Colonel R. S. J. Thirwell RFD ED, as Treasurer.[338]

The draft DRA constitution was adopted at this meeting. Somewhat curiously, while State committees were allowed 'no more than three vice-presidents to allow Tri-Service representation', this was not the case in the National Executive, which only allowed for one, with no provision for any tri-service roles at that level. Remarkably, it was not until 2011 that the first tri-service vice-presidents were elected at the national level, almost 20 years after the Association decided to become a tri-service organisation. Of interest was that all the Branch submissions to the AGM were 'Army' in orientation, or at least had been prepared by ARes branch presidents. The Victorian president's annual report of September 1993 was a case in point. Other than a passing reference to dialogue with 'the Services' the tri-service nature of the DRA saw no reference at all.

DRA Branches were trying to extend membership to Navy and Air Force reservists, but with varying success. NSW branch already had an Air Force officer as vice-president thanks to the earlier amalgamation with The Reserves Association. Victoria branch was soon to add a Navy Reservist to its committee and Tasmania has successfully incorporated other Services into its membership for some time. With the small numbers of Navy and Air Force reservists in some States, other branches struggled to do so.

Western Australia noted in its newsletter in September 1994 for example that it had continued during the year to extend membership to Navy and Air Force and discussions were held to have an Air Force representative join the committee, but without immediate success—'The President is not having much luck in regard to contacts in either Navy or Air Force'.[339] However, the October 1993 conference was notable for having presentations in the afternoon session not just from ACGS-R, Major General Luttrell, but also from the DRES–AF, Group Captain Alan Barr, and from Captain Kenneth V. Taylor RFD, RANR.[340]

Branch reports at the 1993 AGM showed that the Association had been busy at State level. Membership was growing, and branches were active 'on a number of local issues' affecting the Reserve in their regions, and 'contributed to and supported a number of national matters'.[341] An example of this was the work of branches in responding to the Standish Report with its proposals for Protective Legislation. That report had 'languished for some considerable time awaiting some form of positive action which would ultimately result in the necessary legislation being put to Parliament. It is apparent that those responsible have not viewed the report with sufficient priority....'[342]

Finally, DRA was asked for comments by Major General Nunn. Each branch made comments to Major General Fay who in turn passed them on Nunn for his consideration. Comments on the Standish Report itself contained, not unexpectedly, a range of views. Brigadier Amos said that the document was a 'fairly good attempt to cope with a difficult complex and emotional problem'. Major General Cooke believed the report was 'a typical Canberra-vacuum solution that ignores the reality of human nature'. More carefully, Fay saw it as 'a positive move.... with some practical problems in its implementation.'[343]

The incorporation of the NSW branch under the Companies Act was 'an important achievement' as the Association now had the ability to make submissions to the DFRT (albeit through the NSW branch). The incorporation was an additional benefit from the merger with The Reservists Association.[344] The DFRT confirmed that submissions could be made from November 1993. This, according to Major General Fay, 'caused some consternation' with HQ ADF seeking elaboration of the Associations' intentions. Fay noted to Colonel Sandow (who was to replace him in April 1994 as national president), that 'I have made our intentions/policies quite clear'.[345]

Meanwhile, in comments about an article in the press 'Will Reserve Fight?', Victorian branch members agreed that the RRES should not be criticised as that could lead to diminished assets in the GRES. One senior member said: 'A cynical view might be that when the RRES is no longer fashionable it could be dumped without replacement thereby reducing the defence budget substantially. It was also noted that the RRES will probably disappear as the economy gets better'. Brigadier Greg Garde, a Queen's Counsel and Commander 4 Brigade commented separately that the RRES scheme was seen as a Defence 'creature' rather than an Army 'creature'.[346]

Concerns about the outcomes of the White Paper were expressed at the Victorian branch as early as February 1994. These concerns included that a Reserve 2 Division would not be required, that with only 22 battalions, brigades would be reduced, an expansion base would no longer be required, individual university regiments would not be needed, and training groups would be dismantled. These fears were to be realised. The fact that so much was known so early about this White Paper, indeed rumours had been circulating a year earlier, was perhaps suggestive that deliberate 'leaks' were being made to test reactions and resistance to its projected outcomes. Colonel Sandow, president of the Victorian branch (he was succeeded by Colonel Michael Vincent on 1 April) commented; 'Depot closure was being done with stealth and no thought was being given to future requirements.'[347]

The idea put forward by Major General Glenny when Commander 2 Division in February 1994 of 'thirding' (the Thirds Constant or the 30-30-30 notion') was also a subject of discussion. Glenny had postulated the idea to Land Command that to manage a minimum level of operational capability (MLOC), i.e. the manpower, equipment and level of training needed by a General Reserve unit to reach Operational Level of Capability (OLOC) basically, with one third of posted strength qualified and available for collective training; one third in training or supervising training, and one third unavailable for training.[348] The idea was not taken up by Land Command.

The DRA, since the change of name in 1992, had become involved the Kindred Organisations Committee (KOC) which had been brought together under the RSL's leadership around common concerns for the welfare of the Service community.[349] In March 1994 Major General Cullen wrote to the national president of the RSL, Major General William B. 'Digger' James AO MBE MC, to ask if the DRA could be represented on the RSL's Force Structure Review. It was subsequently agreed that Colonel Sandow would be that representative as DRA national president, following his succession of Major General Fay who had stood down that same month. As part of that shift, the National Executive moved to

Victoria following the AGM in October 1994.[350] There was also change in NSW, with Major General Sharp standing down as State president, to be replaced by Major General Glenny.[351]

In another development in March 1994, Major General Luttrell, ACDF-Res, endorsed the findings of the Army Report on 'The Protection Required for the Civilian Interests of Reservists When Called Out'. In his endorsement he noted that the DRA saw the proposals in the report as 'well balanced, reasonable and a positive step, covering all of the key issues which had been concerning them on this issue'.[352] Just a month before, when Major General Nunn was still ACDF-Res, the DRA had provided its comments from its branches in response to the Standish Report. While Luttrell had echoed the sentiments of that covering letter from the then DRA President, Major General Fay, which concluded that 'the proposal is a positive move and covers all the main issues'. Luttrell's endorsement did not add the qualifier 'We do believe it has, however, some practical problems in its implementation.'[353]

The Standish Report had taken more than six years to come to fruition. The extended Call Out Legislation enacted in July 1988 and the subsequent Army Working Party to examine the issue resulted in a draft report in January 1991. In turn a Review Team report was not completed until November 1993. Final recommendations were endorsed by ACDF-Res in March 1994. The wheels turned slowly indeed, and this was just one of the issues that the DRA and others such as the Defence Reserves Support Committee (DRSC) had been trying to influence and resolve. Suspicions remained; the Victorian branch was later to comment: 'There are problems with the Call Out Legislation. That which is currently in preparation is both draconian and mean. The message being conveyed is 'don't join the Reserve. It is too risky.'[354]

In his first report as national president in May 1994, Colonel Sandow noted that 'the operation of the Ready Reserve (RRes) has left a good deal to be desired' and that ex-Regular and General Reserve personnel have not sought to join the RRes 'as anticipated'. He said that he had already 'gained an impression that that a substantial curtailment of the RRes may be in the pipeline.'[355] He also foreshadowed rumours that there were to be the changes to Training Command with subsequent impact on Reserve training and the role of university regiments. By July 1994, the DRA had become thoroughly alarmed that the Army was proceeding secretly with plans—under Project WELLESLEY—to determine the future role and structure of Training Groups.

Branches from the smaller States in particular, were fearful that rationalisations in training groups and university regiments would result in drastic cuts in the number of senior appointments available. In letters to Major General Luttrell and to Major General Stephen Golding RFD, ACGS-R, Colonel Sandow asked: 'Does this mean that Army are planning to weather a storm of protest as a result of preparation and sudden release of secret plans?'.[356] That Sandow would use such dramatic language was an indication of the lack of consultation and information from Army about these issues. It was, however, anticipated that by the end of 1994 several strategic papers would become available which would contain guidance for the development of the Reserve Component of the Army. For example, the Government's White Paper—*Australian Defence 1994*, and the Army Strategic Plan.

That guidance was expected to develop three main thrusts for the Reserve: to improve capability within known constraints; to strengthen availability of reservists for training and call-out; and to further develop integration of Regular, Reserve and civilian components of the Total Army.[357] There was anticipation, but the concerns expressed by Colonel Sandow would not go away with the release of the reports; if anything, they were heightened. Indeed, Sandow was later to say that 'the emphasis of the White Paper is that the Reserve is nothing more than a part-time Regular force whereas our concept has been along the lines of *Citizen Soldiers*. The 'Come as you are' war seems to be becoming a mindset within Army and Defence in general. We need to challenge the concept of just what a Reserve Force should be.'[358]

Major General Cullen, after attending a briefing and dinner with the CGS, Lieutenant General Grey on 25 August 1994, wrote a letter of thanks. By now 85 years old, Cullen was, however, always happy to take any opportunity to support his beloved Reserves. In his letter to Grey, Cullen took up the cudgels to plead for OCTUs in each State rather than putting all training at RMC Duntroon, as well as supporting the rationale and effectiveness of university regiment training for undergraduates. Cullen accepted Grey's assertion that the One Army concept had been accepted and was now normal but warned him nonetheless not to 'throw the baby out with the bathwater' and 'not to take logic to its illogical conclusion'. Cullen's letter was above all, polite and respectful, but it also showed he had lost none of his enthusiasm for making a point or two to remind the Regulars that they did not always get it right.

The NSW branch newsletter of October 1994 noted that from the branch perspective, the three broad areas of interest were the size, resources and opportunities of the Reserve, in-service terms and conditions and post-service entitlements. It also noted that 'the best representation occurs with a single, National Association voice articulating well-researched and developed standpoints in the proper fora.'[359] Major General Sharp also commented in the newsletter that the MCS was now law and 'creates schisms within the ADF'.

However, he also reported that two aspects were successfully implemented 'in our favour'; the adverse aspects were not applied retrospectively and also that an ex-Service member would become a permanent member of the Appeals Board to be set up under the MCS. He noted that these successes were not due to DRA efforts alone but as part of the efforts of other 'companion organisations'.[360] Still, in a separate note to Colonel Sandow covering the NSW president's annual report Sharp said: 'There remains plenty of concern here as to whether DRA is yet being heard enough in the corridors....'[361]

That same month Sandow wrote a letter of congratulations to Peter K. Reith AM, MHR (Liberal/Flinders) on his appointment as Shadow Defence Minister for the Opposition. The letter was pessimistic, frustrated, almost disillusioned in tone. 'For the past six years or so we have been watching the palpable deterioration of our Defence Forces ...', Sandow began, stating that 'Deterioration of the Army Reserve in particular is becoming a self-fulfilling prophesy due to starvation and patronisation.... our efforts to obtain constructive comment from either political party have so far been virtually fruitless'.

Calling the Ready Reserve expensive and 'disappointedly unsuccessful', Sandow noted that the Reserves were being 'curtailed with respect to finance, equipment, facilities and senior rank positions.'[362] In that respect Sandow noted that that were 'less than one quarter of the number of senior officers in the Reserve as there are in the Regular Services....The plan to create a Joint Services Operational Headquarters [will add] yet another strata of rank and command at the top of our tiny defence system.'[363] Several months later, in February 1995, the Government announced a review of the Ready Reserve would be conducted by Lieutenant General H. John ('John') Coates AC MBE and Dr Hugh Smith, Department of Politics at the Australian Defence Force Academy (ADFA). The review was required to make an overall assessment of the scheme and recommend modifications as appropriate. Was the review a response, at least in part, to the DRA criticism of the scheme?

The DRA National Executive in Melbourne was starting to meet monthly by 1994. The surviving minutes of August through September of that year give some idea of the work underway, albeit with only three in attendance—Colonel Sandow and Lieutenant Colonels John F. Henry and Ian George. Correspondence in and out was waded through covering a range of matters from the pressing to the mundane, while the details of the national conference scheduled for October was discussed in detail as arrangements for venue and speakers was well underway. There was nothing dramatic about this work—the day-to-day minutiae of a national association, but it was a reminder of the entirely routine but essential work behind the scenes of not just the National Executive of the Association but also the work going on in committee meetings and arising from the contributions of ordinary members in State branches right around the country.

The contribution of those office-holders at all levels of the Association since its inception, often forgotten or unrecorded as time goes by, can often be overlooked when national issues are at stake. But those contributions were the 'grist to the Association's mill' as it determinedly followed its agendas in support of the Reserves through the many years of its existence. For example, Tasmanian branch reported 335 members of which 250 were financial that year, and NSW branch reported 149 ordinary members and 66 life members. Numbers such as these were the very lifeblood of the Associations existence.[364] There had always been a tension between those who felt that more members were essential to give the DRA more *gravitas*, and those who felt that a small coteries of senior experts should be the 'brains trust and engine' of the Association, but the reality was that Association numbers and contributions were not linked, and both waxed and waned throughout the Association's existence.

The DRA AGM and conference was held at Randwick on 4 November 1994, with the keynote address by Michael J. O'Connor, executive director of the Australian Defence Association (ADA) with which the Association had had cordial relations with since the late 1970s. The agenda for the conference included topics such as the constitution, MCS Report, Defence White Paper (the White Paper went to the Senate that month), Future of Training Command and the 'Manifesto'. The 'Manifesto' was developed by Colonel Sandow as an initiative to define the DRA aims and objectives and clarify primary issues for focus. By September 1994 a draft was already being circulated in branches for comment and suggestions. That process was to continue through to the AGM and conference of October 1995.

Commander Peter A. Hardy, Assistant DG—Reserves RAN addressed the Queensland branch of the DRA in February 1995. In a wide-ranging presentation one key point was made: 'Navy is attempting to generate an ethos that the only distinction between full time and part time members is the time on the job'. He went on to say: 'The separation of the Permanent Naval Force (PNF) and Australia Naval Reserve (ANR) cultures through forty years lead to negative stereotyping in both camps'. However, he said, 'there is a need, even with integration, to maintain a distinct identity for the ANR'.

Noting that the ANR was 'rapidly moving from a force partially trained for war in peacetime to a force which was contributing to the 'sharp-end' of day-to-day operations on a part-time basis', Hardy compared the ANR to the stock market, 'with daily fluctuations but an upward curve which will provide a very healthy dividend in the long term'.[365] His paper noted that the ANR's three components consisted of the Naval Ready Reserve (155), General Reserve (1300) and the Standby Reserve (3500).[366]

Work on the 'Manifesto' continued with Colonel Sandow noting at a meeting of the Victorian branch in April 1995 that 'For all causes we need hard facts rather than anecdotal evidence if the Association is to be taken seriously'.[367] In May Major General Glenny was asked to do more work on the section dealing with finances. Later in November 1996, other senior members of the association were asking what had happened to the 'Manifesto'—was it complete?[368] Work on the important paper for the Association was indeed continuing and it would be resolved by early 1996 in a major policy statement by the Association. Meanwhile, in NSW, Glenny replaced Major General Sharp as branch president—Glenny would also be elected vice-president of the Association at the next DRA AGM in October.

At a seminar held by NSW branch in May 1995, it was addressed by Major General Brian A. McGrath RFD, Commander 2 Division who presented on the significant issues affecting the Reserves. The Reserves had been recognised for the first time as an integral part of the ADF structure, said McGrath, but recruiting and retention remained the major problem area for the ARes.[369] A 'critical review' was in progress, due in September. McGrath was citing the Army Manpower Control Study—Stage 3 ('Jans Study') by Brigadier Jans with its focus on retention; completion date for Army Office was delayed to December. DRA president Brigadier Sandow was of the belief that 'Army will listen to them [Dr Jans] rather than us'. Nothing more was heard about the report.[370]

Queensland branch also conducted 'workshops' for its members to investigate and discuss current issues around service and administration conditions for ARes members; these were valuable exercises to throw up new ideas and perspectives for the National Executive and council to consider.[371] DRA members continued to complain that despite a July 1995 deployment of Reserve doctors to Rwanda for 6 weeks, the Reserves received no recognition of this.

The DRA national committee was at work on three submissions—on the Defence Reserve Call Out Legislation, the Ready Reserve and Project WELLESLEY. Meanwhile, even as the implications for Reserve training and university regiments began to sink in, a new and even more comprehensive review was underway.[372] In October 1995, 60 ARES officers met at RAAF Amberley in Queensland to give input into a major review by Brigadier Peter J. Dunn ARA, looking at the 'Army in the 21st Century'. Known as 'A21', the review aimed at identifying the most appropriate structure of the Army of the future and was 'the most comprehensive review of the Army's structure for many years'. A21 was to become the most controversial review in its impact on Reserve forces since the 1987 Force Structure Review, and over time the senior leadership of the DRA and its members would once again come together in resistance to its eventual policy changes.

The DRA AGM and national conference was held in Melbourne in late October 1995. The CDF, General John S. Baker AC DSM gave the keynote address. Outstanding among the branch reports was Tasmania, which reported 620 members. Not all were so robust—Queensland branch president Lieutenant Colonel George L. Hulse also reported in his annual summary that the branch had 50 members but needed 60 to remain financially viable. He noted the complete lack of any Ready Reserve as members.[373]

Reporting on the conference early the following year, the NSW branch newsletter recorded branch president Major General Glenny's scepticism that despite indications from [incoming Liberal/National] Government and Defence interest in and support of the General Reserve, it would be measured against failures to actually do anything much thus far:

> The 1972-74 Millar and other studies recommended a series of measures to enhance recruiting and retention—'the General Reserves 22 years on are still waiting'. Compare this to 'the treatment of the Ready Reserve which saw within two years overly generous financial incentives in place when many General Reserve units and individuals are at shorter degrees of readiness.'
>
> Legislation to call-out reservists was introduced in 1988 but the protection of Civil Interests of Reservists on call-out is still awaiting legislation 8 years later, and 'may be another 2-3 years to become reality'. Reservists and the DRA support the concept and practice of 'One Defence Force'...but such support does not extend to the total absorption and immersion to the degree of NO recognition.[374]

The newsletter noted that the DRA conference was conscious of the statement made on 22 January 1996 by Archibald ('Arch') R. Bevis MHR (Labor/Brisbane), Parliamentary Secretary to the Minister for Defence, entitled 'New Reserve Directions', and would see if the proposals would be implemented. The Bevis paper quoted the CDF, General Baker, who said that Reserves 'are a key element of our concepts now, they are not a second eleven. They are part of the Total Force and will be regarded as such and trained and equipped as such'. Citing the [Labor] Government's commitment to the importance of the Reserves, Bevis stated that the Reserves 'will receive our support into the 21st century.'[375]

Colonel Sandow wrote to Bevis no doubt questioning that commitment, to receive an 'absurd answer'. Sandow commented to Major General Cooke in correspondence to say: 'How can politicians imagine that their platitudes can be in the slightest convincing to intelligent people?'. He proposed a 'stiff reply'.[376] There was no need as Labor was swept out of office soon after. No doubt this outcome was to the satisfaction of many reservists who had long bemoaned the fact that the Reserve always seemed to be on the receiving end of the more negative outcomes of the endless reports and reviews and white papers over the years. It remained to be seen whether the new Howard Government could overcome the cynicism evident among some members of the DRA about politicians and Defence bureaucrats, both civilian and uniformed.

With circulation of the draft 'Manifesto' through branches in 1995, giving rise to a range of variations around different aspects, it was agreed at the national council meeting in October 1995 that a discussion group was needed to resolve the outstanding Association issues regarding DRA national policy. Accordingly, in February 1996 a two-day conference of DRA delegates from all State branches in Sydney was held for the purpose. The 'Manifesto' was subsequently agreed upon—with the rather ponderous title of 'Defence Reserves Association Statement of Policies in Respect of Reserves in the Australian Defence Force'. Lieutenant Colonel John S. Haynes OAM also submitted a paper at the meeting on 'Service Conditions and Industrial Matters', which was to be considered and later incorporated into the 'Manifesto.'[377]

Commander Malcolm J. Hedges RFD RANR furnished committee members with a copy of Ch7 [Reserves] of the 'Glenn Review', or more formally, the 'ADF in the Twenty-First Century (Personnel Policy Strategy Review): ... 'there appears to be a push with ADF to redefine what a Reservist is and, in this way, may restrict the rights of citizens to serve their country in the traditional manner.'[378] The Glenn Review itself stated:

> The Committee received an abundance of evidence to show that, over the past 20 years or so, there have been many sound ideas about how the performance and conditions of the ADF Reserves could be improved. No attempt has been made by Defence to implement those initiatives in a sustained and cohesive way. Fragmented policies, inadequate priority for the resources available and lack of commitment have been the hallmarks of Reserve development.[379]

The Victorian branch like many others were watching developments with alarm:

> One of the functions of this Association is to support the right of the citizen to give Reserve service. With the push towards FT/PT service in the ADF we consider this right is being eroded. The organisation being developed [from Project WELLESLEY] seems to be undermining this.[380]

In April 1996 Major General Glenny was invited to speak to the Liberal Party NSW Defence & Foreign Affairs Committee. It was an invitation that he took upon with enthusiasm, given that the Liberal/National Coalition had just powered back into Government with a 29 seat swing towards it. He took the time to 'educate' his audience about how the three Service reserve components worked but he also reminded them of the promises made by the incoming Government. Glenny summarised the call-out protections of reservist's civil interests and noted:

> Despite working parties, reviews, endorsements and promises, legislation is still not in place 9 years after the announcement of change and 8 years after Reserve obligations were changed, to cover the promised Protection of Civil Interests...Most general reservists and their supporting associations see this as another example of performance not matching promises and contrast the 8 year delay against the almost immediate introduction of substantial benefits and conditions of the Ready Reserve.[381]

Glenny concluded by saying: 'Defence is too important to be left to the Defence department politicians and generals. We need vigorous, ongoing and well-informed defence debate'.[382] Colonel Sandow, DRA president, meanwhile was discussing with Lieutenant Colonel Hulse a DRA project proposed by Queensland branch around Reserve fitness and health for individual readiness. In the context Sandow made some interesting comments on F/T and P/T members of the ADF, asking Hulse to be conscious of, and not to indicate DRA acceptance of:

> ...the proposal developing amongst senior Regular and Reserve officers that there is no need for a 'Reserve' but rather there should be [F/T] and [P/T] members of the Defence Forces. The notion is a high ideal which has much appeal, but it needs a lot more thinking through than to be accepted as a fashionable concept. The trend of 'part-time employees' may eventually be supported by logic and common sense and found to be practicable (although I cannot at this stage envisage it) but it is not current policy of DRA.[383]

8.
A21 Alarms

On 22 May 1996, Major Generals Cullen and Glenny attended a meeting with the Minister for Defence, ostensibly a courtesy call to invite the Minister to address a meeting of the DRA NSW members and other interested defence members later in the year, but most likely to discuss the A21 policy development affecting the Reserve.[384] Shortly afterwards the DRA National Executive finally issued to its branches a 'Statement of Policies in Respect of Reserves in the Australian Defence Forces'. Coordinated by Major General Sharp, and approved by all branches, the document set out the aims of the DRA and stated: 'The Association has moved to a more proactive role by providing a structure of policies from which it develops consistent approaches to improving the position of Defence reserves in the overall scheme of the defence of Australia'.

It then set out those policy areas as follows:
1. Reserves in the force structure—the DRA set out that Army Reserves should comprise 50,000 (against an Army strength of 75,000) with RAN and RAAF Reserves both set at 8000.
2. Service and post-service conditions including terms of engagement, injury compensation and superannuation.
3. Training and development including general training, skills development, officer training and career development.
4. Honours and awards
5. Cadets
6. Publicity[385].

The Tasmanian branch was tasked by the National Executive to produce a paper on retention.[386] NSW branch would develop a paper on superannuation and compensation.

A later report showed that by this time, 44% of the Naval Reserve and 60% of the RAAF Reserve were ex-regulars.[387] To many of the Army Reserve members of the DRA, the vast majority, these figures seemed to portent a similar situation would arise in the Army Reserves. How this would change the character and traditional ethos of the Reserves was no doubt a discussion of concern in retired Reservist officer circles. However, as Major General Glenny had pointed out in his address to the NSW Liberals in April, RAAF and Navy Reserves normally had higher components of ex-regulars in their components because of the way they used individuals, whereas in Army 50% of its strength and 60% of its operational units were Reserves, with a smaller ex-Regular component of between 5-10%.[388]

Nonetheless, ongoing media coverage for and against the Reserves and the issues continued. The *Australian* defence writer Don Greeles wrote a front-page leader on 1 July 1996, criticising the Government for training cutbacks in 2 Division to make budgets, leading to one of its brigades 'being stood down.'[389] An unnamed Reserve officer was quoted as saying that the suspension of all training in [4 Brigade] had undermined morale and reduced the effectiveness of individual units.

The next day Michael O'Connor, the president of the Australian Defence Association also wrote an opinion piece in the *Australian* about the 'long running struggle within defence and especially the army over what part the reserves should play in national defence. It is a question', he wrote, 'that has bedevilled Australian policy since 1945 and we seem no closer to a credible answer.'[390] O'Connor noted, that 'when it came to doling out resources, the reserves always seem to come in at the tail of the queue.'[391]

He described the Ready Reserve as 'a classic example of the camel being a horse designed by committee', which a review by Major General Coates and Dr Smith 'concluded in rather unenthusiastic terms that the scheme was useful...but could do better'. O'Connor wrote, 'So severe has been the neglect of the reserve forces that those who are naturally suspicious of official motives tend to see it as a means of quietly making the reserve problem go away by default.'[392] This was certainly the attitude of many members of the DRA and its leadership at the time. Counterattacks on the Reserve and DRA by default quickly came with letters to the editor proclaiming that senior Reserve officers were only interested in promotion and not in the ordinary soldier or young officer, with the effectiveness of the Reserves 'limited by the reservists' own greed.'[393]

By mid-1996 the DRA was becoming increasingly alarmed at the 'Army in the 21st Century (A21)' development. Major General Glenny was among several senior DRA officers at the forefront of concern on the issue. Following his meeting with the Minister in May along with

Major General Cullen, Glenny wrote to Colonel Sandow expressing his opinions at a range of matters adversely affecting the Reserves, both actual and potential. Pent-up frustration at lack of knowledge and possible consequences of the A21 concept was loud and clear in a confidential memo from DRA president Sandow in reply and copied to Cullen in June 1996. Sandow's exasperation could not have been clearer.

He wrote: 'On the matter of A21, the Army and ADF's Office are hell-bent on secretly and subversively planning to nobble the Reserves'. It was not only the Army Reserve, but Navy as well, he contended, noting 'talk of not commissioning sea-going reservists unless they can serve two years full-time'. He continued in the same vein:

> The idea that Reserves can perform as 'part-time' employees is simply nonsense. There are fanciful ideas that they can move between their civilian occupation and bursts of full-time service. On the other hand, it is obviously such a transparently ridiculous idea that there is room to believe that it is being designed for the express purpose of ensuring that it will not work, and that Reserves will, in consequence, be discredited.[394]

Sandow was plainly furious as he detailed the abuse of the Reserves over the years until current times. 'The secrecy of it is in line with the secrecy of Project WELLESLEY, and it is understandable that it is being supported by the same set of puppets', he wrote. He criticised serving senior Reserve officers for accepting and supporting the 'insidious position of the Services regarding the call-out provisions' and 'taking in the whole A21 scheme hook, line and sinker'. 'How on earth can they accept the manipulation of the Reserve into an impossible position?', he asked.

Agreeing with Glenny's positions, Sandow said:

> There is not 'One Army' and those that loudly profess it are exactly those who are most working to ensure that it will never happen…You and I know that the Regular and Reservist are two different beings. Each has enviable characteristics. Neither can hope to emulate the other. When the time comes, the one that can save the day is the Reservist.

Sandow hoped the new Minister for Defence could yet be educated to understand the Reserve position. 'The regular army should be regarded as the caretaker of the Reserve—there are not enough of them to be anything else', he wrote.[395]

It was not surprising that Sandow had made the memo 'confidential'. Putting aside his direct criticism of senior Reserve officers by name, 'towing the party line to become ACDF'—he also made egregious comments about the Regular senior officers as well: 'They have too many dull people at senior levels. I know there are a few bright ones', he acknowledged, 'but collectively they are too dull to inspire the Reserve, and now we have them determined to create a Reserve hierarchy which is also too dull to inspire the Reserve'. Sandow noted that the

Minister for Defence had accepted an invitation to attend the DRA annual conference and he, Sandow, had begun to prepare his opening address and 'expect to make it direct and frank'.[396]

This private exchange of correspondence was followed up by Major General Glenny, as branch president, in the NSW branch newsletter in August 1996 with a long and passionate message to the members. It was becoming clear to many reservists that the One Army concept was also meaning no recognition at all for the General Reserve. The example of Reserve medical team contributions to the Rwanda deployment was a case in point; there was no recognition that Reserves were there at all. This was not what the DRA had agreed to when it decided to support the concept.

When the Liberal–National Coalition had come to power in 1996 it had implemented its promise to abolish the Ready Reserve and make a commitment to the General Reserve. A21 would test that commitment, said Glenny:

> We are again seeing dramatic change preceded by the type of statements made at the time of the Army Reserve Review of 1987/88 and the Force Structure Review of 1990/91…I leave readers to judge whether the loss of the 3rd Division, four Reserve and two Regular battalions along with numerous supporting units which were either disbanded or amalgamated, have produced any of the benefits that were put forward to justify the change.[397]

Glenny highlighted his concerns with numerous examples. He questioned how renaming of divisional and brigade headquarters and formations added to operational capability; and whether elimination of units and formations just led to bigger headquarters in or near Canberra. He described how command and staff relationships and trust between Regular and Reserve components was breaking down due to Defence secrecy and 'need to know'. He pointed out how reservists could not generally fill F/T positions if they want to command, leading to those available and not the best being appointed, and asking what meaningful, acceptable progression existed beyond the rank of major?[398]

Glenny continued, 'Did A21 only mean support and implementation of Reserves as fillers in unit establishments?'. He noted training days despite unmet promises to increase them remained at the same 33 days per member on strength, which had been reduced from 55 days in 1968, 18 years before. And this after 40 plus reports and inquiries over that period along with three major reductions in units and formations in the past eight years alone. He said:

> Integration, while commendable, is dangerous if Regulars train days and Reserves train nights and weekends. It must be acknowledged and practised with conferences, exercises and training times which recognise that protection of reservists' civilian occupations are their primary concern.[399]

Glenny's conclusions were gloomy. 'The total reorganisation of Army, with the probable loss of corps, formations and units against a background of secrecy and no vigorous, widespread debate, is frightening', he wrote. He stated that 'the creation of further one-on-one strategic headquarters whilst eliminating headquarters at the operational level to give the appearance of a flatter structure' was inimical to the development of an effective and efficient defence force, as was '...constant change and turmoil with no changes lasting long enough to have its objectives tested before it is overtaken with something else'.

Noting that 'One Army' was a reality, Glenny also said that the One Army deserved better than A21—as did Australia. He asked that previous promises for the Reserve become reality before the Reserve is again challenged by [A21]. Glenny's message in the branch newsletter—which he said was his personal viewpoint—was powerful, a cry for sanity and common sense, an appeal for Defence and the Government to 'recognise the skills, characteristics and strengths that Regular and Reserve components bring to the Total Force.'[400]

For at least 18 months the various commanders had been giving assurances of the way forward—Major General McGrath, Commander 2 Division in mid-1995, Lieutenant General Baker, CDF and Lieutenant General Sanderson AC, Chief of Army (formerly CGS) in 1996—all had identified and lauded a significant future role for reservists in future ADF operations and better utilisation of the Reserve. However, Sanderson did add that employment details and what would happen to the 26,000 strong Army Reserve was 'still being worked out'. Glenny's message clearly set out the DRA's position—'in acknowledging that the increased use of and reliance on the Reserve component is fundamental to the 'one force' concept, the objective could not be met to the detriment of the civilian careers of the individuals concerned'.[401]

Major changes were also coming for the RAAF Reserve. The Director of Reserves—Air Force (DRES-AF) had been tasked by the Chief of Air Staff (CAS) as far back as 1995 to produce a paper for the CAS Advisory Committee on restructuring the Air Force Reserve. The paper, 'A New Role for Reserve Forces in the RAAF' was accepted on 1 September of that year. The ensuing reorganisation was designed to fully integrate its Active Reserve component with the Permanent Air Force (PAF) through postings of reservists to units which became responsible for their training, employment, and readiness.[402] *RAAF News* of April 1996 however, noted that the proposed restructure of the Air Force Reserve would need a number of legislative changes (there were 15 different Acts and nine separate Regulations affecting the RAAF Reserve), which were unlikely to be in place before 1998.

Changes in administrative command were underway but again a review of the Reserve Squadrons in late 1996/early 1997 had to be conducted first. Commenting on these changes, within the proposed reduction of RAAF numbers from 22,500 to 14,500 uniformed personnel, Squadron Leader Gilbertson, the NSW branch vice-president noted that '...the fact remains that in the current uncertain climate, many Reserve members are left wondering just what the future holds for them, in their 'new and enhanced roles as professionals'. As Gilbertson said, '...one has to seriously question how successful the 'part-time professional' concept will be'. In news from Navy, off-the-street recruiting had resumed for reservists to commit to two years' service within the Ready Reserve Scheme in 1995. Navy downsizing was not as significant as Air Force, but still had 799 billets being tested for 'civilianisation'. More sea-going billets, however, were expected to open up to Reserves.[403]

In early September 1996 A21 was again in the DRA sights. This time it was Major General Cullen who wrote to the Minister for Defence Ian M. McLachlan MHR (Liberal/Barker) following a briefing—without any documents—at Defence by the Land Commander Australia, Major General Francis ('Frank') J. Hickling AO CSC. The briefing was held for Cullen among other retired Reserve generals on the A21 scheme. As usual, Cullen came straight to the point: 'To say I was astounded and distressed would be an understatement', he wrote, '...I have to say they are trying to make omelettes without eggs'.

After Lieutenant General Sanderson, Chief of Army, also added to the briefing, Cullen said:

> ...if I closed my eyes, it was just the same motherhood words and clichés that Lieutenant General Sir Reggie Pollard, the CGS at the time, used when 'selling' the Pentropic Division Reorganisation Plan. It took Australia 2 or 3 years to get rid of that nonsense with its several similar disastrous aspects. Thank goodness...that the Ready Reserve has gone or else it would be even more difficult for all concerned to abolish the A21 Plan.[404]

Cullen asked the Minister why he could not arrange an 'unofficial' conference chaired by the Minister around a table with Generals Baker and Sanderson and Brigadiers Dunn and Sandow, Major General Glenny and himself for 3-4 hours? He added: 'The serving Reserve Generals are too disciplined and subservient to be frank.'[405]

In a letter to Major General Hickling in September 1996, Major General Sharp also weighed in regarding the presentation to the 'experienced, but non-current, [DRA] group' from Hickling around A21. Sharp described it as 'a gagging, secretive, non-transparent and sometimes hurried' presentation, which 'sounds very loud alarm bells'. He also asked a number of pertinent questions about the A21 approach, not least to query why the *Citizen Soldier* was once again being neglected or ignored in A21. He went on to say...'the anti-*Citizen Soldier* mindset, both deliberate and unwitting, is very apparent in the Army in 1996'.[406]

Correspondence continued to flow between the DRA and Branch presidents with the Minister for Defence over the A21 concept. In September 1996 a letter from Peter Jennings, the Minister's senior advisor, was received in reply to a letter of concern from NSW president Major General Glenny. In that reply, Jennings reiterated the Government's support for the Reserves, while noting that the implementation plan for A21 had yet to be endorsed and that 'final restructuring would depend on trials to be conducted over coming years'.[407] It is not clear whether these reassurances of 'increasing Regular personnel in Reserve units' to help Reserves meet their commitments was what Glenny was looking for.

The DRA AGM held in Melbourne 19-20 October 1996.[408] It was addressed by the Minister for Defence Ian McLachlan and the CGS Lieutenant General Sanderson. McLachlan focused on 'Future Directions' including the restructure proposal for Army and the DER, which Parliament had announced only a few days before the conference. He also confirmed that defence spending would be quarantined from across the board budget cuts (under Labor the defence budget had decreased from 2.6% to 1.9%).

McLachlan advised that the Army would trial a Task Force structure—partly to get away from the previous Government's 'defence of Australia' syndrome—with a focus on Victoria. The Minister noted that more personnel including from Reserve, would be moved from administrative or support tasks to operational.... with more emphasis on Full Time Duty [FTD]—which also would allow Regular personnel to work with Reserves. Lieutenant General Sanderson in his overview, highlighted the need for Army to move towards total integration as had the Navy and Air Force during the late 1980s and early 1990s.[409]

Presentations from the DRES-AF, Group Captain Anthony ('Tony') P. Behm, and Director Navy Reserves (DNR), Captain K. Taylor, followed. By comparison to Army, their issues seemed positively benign. Their reserve components were a good deal smaller than Army for a start—RAAF Active Reserve had 1000 slots filled of an establishment of 1550 for example. However, they were, as alluded to by the Minister, more advanced than Army in the integration with their Permanent components, not least because of the high level requirement in those services to operate complex fleet and air ships, aircraft, and equipment. Taylor envisioned a new relationship between the Reserves and Permanent Navy in his papers 'Towards a Total Force' and 'Approaching a Total Force'. Taylor's concept for reform became a blueprint for the realignment of the Reserves as an integral component of the Navy.[410]

The presentation by ACRES-A, Major General Golding AM RFD, was accordingly a good deal longer. Golding stated that the ARES needed a greater capability, in turn dependent on availability of PT personnel. Many of the stated means to this objective must have sounded familiar to the older DRA 'veterans', such as 'better focused initial training', better employer support', better career opportunities', etc.[411] Soon after the national conference, the Minister for Defence Industry, Science and Personnel, Bronwyn K. Bishop MHR, announced a 'comprehensive inquiry into military compensation and rehabilitation'. The DRA was not named as an organisation regarded as among those which must be consulted, and which were to be invited to make submissions despite being one of the strongest advocates on the subject for years. The working group was to report by 1 March 1997 through an interdepartmental steering committee.[412]

Later Colonel Sandow wrote again to the Minister for Defence in November on a new issue, namely the appointment of a regular CO for 5/6 RVR. Associated with this was the apparent request for Reserve brigadiers to commit to full time service if appointed to command brigades; this was seen as another way to appoint regular officers instead given that most Reserve officers at that rank were also in senior positions in their civilian jobs.[413] Sandow had been able to see the CGS on the issue but not the Minister. McLachlan replied on 26 November, ending with the usual exhortation: 'Your support in helping to ensure we develop a more capable Reserve component would greatly assist this very necessary restructuring exercise'.

Sandow saw this as inferring that the DRA was not in fact supportive and wrote back accordingly. Sandow began by noting that 'We are simply trying to convince you that there are two sides to this question and that our perspective on the subject results from decades of intimacy with the problems.'[414] He continued:

> It has not been our intention to inflame the core of prejudice about Reserves, but simply to seize the moment of a new, openminded administration to establish a few reasonable ground rules in a climate of the regression which has continued for many years. There is an extended catalogue of anomalies which, if permitted to continue, will ensure a miserable harvest no matter how good the present intentions are. Planning and implementation so far have indeed been devious, and the content of the briefings has incomplete and largely irrelevant.[415]

In mid-October 1996 the Minister for Defence established the Defence Efficiency Review (DER, or what became known as the 'Macintosh Committee') to make recommendations for reforming Defence management and financial processes to 'produce the most efficient and effective Defence Force within current budgetary constraints.'[416] The DRA Victoria branch asked: 'Was there a need to make a submission to the Macintosh Committee? ...some contribution was needed 'if only to show our presence.'[417] It was agreed the National Executive would make a submission.

Once again, the promotion of a regular officer, Lieutenant Colonel Michael Godfrey, over four available, qualified reservists as the new CO of 5/6 RVR caused angst; Colonel Sandow raising the issue with the Defence Minister, Ian McLachlan.[418] Major General Cullen also entered the fray writing to McLachlan as well on 19 December. The letter took aim at General Sanderson in particular. Cullen then went to the Prime Minister asking, 'Do we really want to eliminate the career path of the potential brilliant part-time officers who cannot afford to become full-time commanders in peace?'.[419]

By December 1996 the state of the Reserves had reached such a point that Major General Cullen, on behalf of Sir Roden Cutler VC AK KCMG KCVO CBE, and Major Generals Broadbent, Glenny, Maitland, Murchison, and Sharp felt compelled to write an open letter to the Prime Minister, John W. Howard, (Liberal/Bennelong). 'A crisis has occurred about the continued existence of the Army Reserve as a contingent force', wrote Cullen, requesting an urgent discussion with the PM within the following two weeks, noting that the DRA and himself personally had been in 'constant communication' with the Minister for Defence, 'but with little effect'. The main point was the assertion that 'under the guise of A21 or the reorganisation of the 'Total Army' or the 'One Army' the Army Reserve is being destroyed as an effective factor in the Defence of Australia'.

Cullen asked, 'Do we really want to eliminate the Citizen Force?'.[420] In particular the generals felt that by pursuing the present policy of the CGS (Lieutenant General Sanderson) and 'what's worse, his practices and decisions, that destruction [of the Reserve] is being perpetuated'. Unfortunately, Sanderson's recommendations are being adopted, said Cullen, noting that the decision re 5/6 Battalion RVR 'was the current crisis but is also the tip of the iceberg.'[421]

Sanderson addressed the RUSI in Sydney on 18 February 1997 (his title became Chief of Army the following day) and the occasion was attended by many DRA members.[422] The event was important to the DRA's understanding of where Defence was heading, especially Army, on some important Reserve issues. Sanderson was given two questions—what was the future of the Reserve and what has happened to the promised Protection of Civil Interests under Call-Out provisions for reservists for which we have been waiting for 10 years? Sanderson's reply to both questions was illuminating. To the first he replied:

> I touched on the need for dramatic cultural and actual change to our Army; that modern technology, communications and command structures need to be different from the past, and that operations will be conducted across extremes of distance requiring a flexible approach and with a task force unknown in the past. Old structures

> are unsuitable .... Many of the changes will demand that Regulars be there to implement them, once we have settled into A21 in three to five years, then Reserve Officers should be again able to fill positions and re-assume command.

The answer would not have filled the DRA officers there with any confidence about the future of the Reserve. In response to the second question, he replied:

> Australians have always volunteered when there is a need. Should a conflict occur in the future, then one of two things would happen. Australians would again volunteer or, if necessary, the Government, recognising the need would enact the Call-Out and individuals and units would be duty bound to respond. Protection of reservist's Civil Interests on Call-Out is therefore low on mine and Army's priorities.[423]

Soon after, the report of the DER—'Future Directions for the Management of Australia's Defence'—was tabled in Parliament on 11 April 1997. The 70 findings and recommendations of the report formed the key objectives for implementation through the Defence Reform Program (DRP). This would become the major issue affecting the Reserves in the coming years, as a new 'Restructuring the Army' program got underway.

Meantime, practical matters in the DRA also need attention. Major General Glenny, president of the NSW branch, had contacted Lieutenant Colonel Nebenzahl, former editor of *Citizen Soldier* magazine in the days of the CMF Association, to inquire whether a national DRA magazine was possible. It was subsequently agreed to produce and release a new national newsletter to be edited by Lieutenant Colonel Robert Sealey, RFD ED. This was a positive step for the DRA, promoting it as a national organisation, but the newsletter would not transpire until 1998.[424]

Meanwhile the replacement of Colonel Sandow as DRA president and move of the National Executive to NSW from Victoria was discussed.[425] Sandow had sounded out Major General Glenny about the possibility of him taking over as far back as December 1996, so it was no real surprise to receive Sandow's formal notification that he would stand down on 1 April 1997.[426] Glenny subsequently became national president. Squadron Leader Gilbertson, although not a 'red tab', became president of the NSW branch and the first State president from a service other than Army, 'until a more senior officer could be approached'.[427]

Although Glenny took over from Sandow in April, he did not have a functioning executive team until September, when Flight Lieutenant Hugh Quelch, who was also president of the IOOF Friendly Society, agreed to become national secretary, but a national treasurer had yet to be appointed.[428] This did not stop Glenny from immediately and actively involving himself in the DRA's issues. On 28 May Glenny, along with Sandow and Major General Cullen met with the PM. Among other things they requested financial support from the Government; this was later rejected by Defence.

Glenny had already circulated his paper 'An Army for the Twenty First century—Fact or Fiction, Perception or Reality', to fifteen influential individuals and DRA State branches, and to the PM. Following the meeting, the paper was distributed by the PM's office to 'ten key military and civil appointments in Defence'.[429] On 17 July Glenny met with ACRES-A to discuss DRA concerns; on the 28th he visited the WA branch; on 15 August he attended a briefing for him and Major Generals Cullen, Maitland and Sharp by the Commander Australian Theatre, Major General James M. Connolly AO CSC. On 23 September Glenny appeared before the Government Members' Defence, Trade and Foreign Affairs Committee and then joined a delegation to the Defence Minister on 29 August.[430]

Glenny's paper, 'An Army for the Twenty First Century' was an impressive and wide-ranging critique. More than 40 pages long, it was carefully considered and brought to bear the full range of Glenny's decades of personal and professional experience with the Reserve and its issues. The paper had also seen contributions from a cross-section of Glenny's peers, and it set out 'concerns believed to be felt by a significant part of the Defence-interested community, concerning the Army's future structure and direction, seen as likely to emerge from the implementation of [A21] and the Defence Efficiency Review (DER) 1997'.[431] Overall, the paper highlighted the failures over decades of successive Governments to deliver on promises made to the Reserve and argued for redress of several of these adverse impacts on the Reserve.

Glenny decried the fact that neither A21 nor the DER was being questioned by the Defence hierarchy any more than previous reviews—there was no debate, no flexibility, and no adjustments—'no one person, committee or review should be taken as holding the Holy Grail and not be subject to wide-ranging and ongoing scrutiny from a wide range of interested parties', he wrote. Above all it was a clarion call for a change in attitudes towards the General Reserve in particular, which Glenny saw as under threat of destruction. It expressed the hope that a review of A21 and DER, even at this late stage, 'may remove the belief amongst reservists that Defence promises are never matched by performance' and not put at risk 'the progress towards One Army so evident during the period 1989-1994'.[432]

In his president's report to the NSW branch AGM on 30 September 1997, Gilbertson said that it was his belief that 'the Association and indeed the [ADF] was at the crossroads'. The Government had accepted almost all recommendations that were contained in the DER and had tasked the ADF and its commanders with implementation through the

Defence Reform Program (DRP). 'Without doubt', he said, 'the DRP would result in the most fundamental change to the ADF since the [Sir Arthur] Tange Review of 1973'.[433]

Coming on top of A21 ('Restructuring the Army' or RTA) on which the Association's views were well documented as a most dramatic change for the Army, the NSW president warned that these changes must impact on both Regular/Permanent and Reserve formations and units, with consequent dramatic changes to ADF doctrine and capacity. Gilbertson said: 'It was his view that bodies such as the DRA must continue to challenge such draconian measures, with their attendant upheaval and effect on Reserve morale'.[434] In conclusion, the NSW president noted that while both Government and Defence were committed to these reforms, and move towards a more operationally-focused ADF, he believed there was still a role for the DRA—that is, 'we must ensure that the Defence Reserves do not get 'left out' of the picture or become the 'the poor cousins' of their regular/permanent/fulltime counterparts'.[435]

The DRA AGM and national conference was held at Watsonia Barracks in Melbourne on 15 November 1997. Reports from the branches showed solid membership—NSW 270, Queensland 59, SA 41, Tasmania 291 and WA 120. A general concern was the potential 'demise of the *Citizen Soldier*', due to retirees from the permanent force being appointed to the Reserve and to Reserve postings. This allied to the need for extensive training under the 'common induction training program', would further diminish the established positions for traditional reservists.[436]

Keynote speaker was Major General John C. Hartley AO, Deputy Chief of Army (DCA) who spoke on the DRP and RTA. The meeting agreed to the idea of a Reserve Forces Day (RFD) march on 1 July 1998, the 50th anniversary of the raising of the CMF after World War II. The RSL, National Servicemen's Association (NSA) and Department of Veterans Affairs (DVA) committed with the DRA to manage the event along with other kindred organisations. Major Generals Glenny and Cullen worked behind the scenes with senior Government ministers including the PM to ensure the RFD concept did not get 'bogged down' in Defence.[437]

In June 1998, the long anticipated national magazine was published—*The Australian Reservist*-edited by Lieutenant Colonel Sealey.[438] It was not very sophisticated, closer to a newsletter with photos, a black and white copy with content dominated by information about the forthcoming Reserve Forces Day on 18 July and excerpts from Major General Glenny's paper 'An Army for the Twenty First Century'. The commercial printer which had initially undertaken the task of printing the initial issue failed to sell any advertising—a reflection of the DRA membership numbers—but the Victorian branch came to the rescue and organised the printing and distribution.[439]

The inside cover provided information on the organisation of the DRA. What stood out at that time was the number of patrons, starting with the Patron in Chief, the Governor-General, Sir William P. Deane AC KBE; followed by a national chairman, Sir Roden Cutler. State patrons were also listed for NSW, SA, WA and Tasmania; only Victoria and the NT did not have one. Whether this was a function of the DRA's central role in the organisation of Reserve Forces Day (which quickly developed its own national organising committees) is not known but it was a trend that would wax and wane over the years. However, combined with the great success of the Reserve Forces Day parades around the country, this was a fillip for the DRA itself, giving a sense of progress at last for those who remembered the old *Citizen Soldier* magazine. By September 1998 Sealey was already pleading for more content for the next edition to come out in early 2000; there was no edition throughout 1999.[440]

Nonetheless, a reading of NSW branch minutes through 1998 show some loss of direction. Incoming NSW president Colonel Graham Fleeton RFD asked members what they saw as the future of the DRA and its activities. Committee minutes reflected a general focus on administration and day-to-day business rather than on some of the large issues the DRA had been involved in in recent years. Discussion revolved around new workshops and issues to discuss but there appeared to be a feeling that the big decisions for the Reserve had all been made, or at least there was nothing more that the DRA could do to affect them.

The involvement in the Reserve Forces Day event seems to have been, on one hand, an opportunity for committee members in all States to get busy with something they *could* control, the organisation of that important event. On the other hand, it was also a major distraction from the big picture of the Reserves, their role, and future issues. A later report noted that 'in providing the time, support and resources to the RFD march in July 1998, the NSW State committee ceased to effectively function between February and July.'[441]

Long serving DRA members could have anticipated that the Reserves would always be facing 'big decisions' and the issues from the DRA president's perspective had certainly not gone away. From 1 July 1998 award simplification provisions of the *Workplace Relations Act* took effect. Leave provisions for reservists to undertake their Defence training was eliminated under the simplified provisions. Shadow Minister for Defence, Arch Bevis was quick to attack the Government: 'At the same time as the Government is restructuring the Army and demanding a greater contribution in training and time from the Reserves[it has stripped] from these very same Reserves the support needed to make that contribution.'[442]

The DRA was not as nimble, given that DRA president Major General Glenny no doubt consulted with some of his branch presidents at least, but he was able to respond in a letter directly to Prime Minister Howard about this and other issues. The Victorian Committee was not one of those consulted—it commented: 'It was agreed that although the thoughts and ideas expressed in the letter accurately mirror those of the DRA generally, there was a need for it to be correctly staffed and framed prior to the release.'[443] In his covering note copying this letter to State presidents and life members of the DRA Glenny noted that: 'The DRA does not question the need for greater access to more training for the Reserve. However, if the increased training is made impossible for employed reservists to attend then the intention was not to improve the Reserve but to destroy it.'[444] His letter to the PM was direct and to the point.

He had previously written to Minister for Defence McLachlan on 28 April regarding dissolution of Reserve bands but had received no reply. However, Glenny noted that the PM had replied to other letters from Major General Cullen and Brigadier Willis, saying in his [Howard's] responses: 'I can assure you that the changes are designed to make the Reserves a more effective component of our military forces'. Glenny accused the PM, the Minister and Defence of 'silence or consent' to allow negative decisions 'which by no stretch of logic other than that displayed by Defence could be said to improve the Reserve.'[445]

Glenny also commented on one news report which stated: 'The Army Reserve is moving towards a crisis with recruiting figures falling to 8 per cent of targets in some States'. The best recruiting figures were from South East Queensland, but it still only managed 51% of its target. Glenny noted that the key factors were common induction training (requiring six weeks full time) and the abolition of Reserve training time from award conditions. Glenny said: 'Reserve training must be available to all skills and professions needed by the Army and not just students...especially since there is now greater reliance on Reserves within the Defence Force.'[446]

Glenny listed the issues for the record:
- The ACDF-Reserves reports not to CDF or VCDF but to Chief of Personnel—and author of A21 ensuring a filtered and distorted input...
- AChief of Army-Reserves. Downgraded to one star and dual posted as Assistant Commander Land Command ('A Reservist holding a dual posting of two of the most demanding postings is effectively neutralised') and no longer on the CA's Advisory Committee.
- Command of 5 and 7 Brigades lost to Reserve.
- Command of 4 Brigade being full-time—few employed reservists are able to obtain the 2-3 years leave necessary.
- Foreshadowed loss of further formations and units.

- Command of 5/6 RVR and 8/7 RVR now Regular -...leads to the desired Defence outcome of NO Reservist beyond the rank of Major becomes self-fulfilling.
- Command to be full-time—impossible for employed reservists.
- Dramatic loss of Reserve officer postings.
- Common Induction Training and Initial Employment Training both to be 6 weeks and completed within 18 months. [Even] Government would not give its employees 12 weeks military leave.
- Removal from awards and Enterprise Agreements of Reserve Leave. Counter-intuitive when Defence is calling for greater reserve availability.
- Protection of reservists' Civil Interests on Call Out. Promised by Defence 12 years ago.
- Elimination of Reserve bands.
- Honours and Awards. A dramatic decrease in awards for reservists across all there services...even when Reserve component is of similar size to the Regular.
- Publicity and acknowledgement of the Reserve forces as part of the Total Force. Hardly any mention of Reserve by Defence or Ministers.[447]

Glenny concluded by saying that despite all of the moderate representations by the DRA around genuine concerns and in specific detail, 'this approach has not seen one issue resolved to the satisfaction of the DRA... [The DRA] is now of the opinion that these concerns will remain unresolved because by design or omission you and the Minister have accepted Defence's intention of 'creating well trained individual reservists, drawn from students and the unemployed, officered and NCO'd by Regulars'. While Glenny regretted the need for the letter, he also noted to the PM that he and the Minister 'can and have rejected all DRA representations' and that they would also acknowledge that the DRA 'has the right to use whatever avenues are available to it to seek resolution of the critical issues it believe face the Defence Force Reserves.'[448]

This was an important letter. The high hopes of improvement that the DRA may have held at the time of the incoming Government after such a long struggle over Reserve issues and frustration dealing with the previous Labor Government had all been dashed within months of the new government coming to power. The stranglehold that Defence bureaucrats and the Regular hierarchy held over the fortunes of the Reserve if anything had been strengthened even further in the past decade. The list of issues, even grievances, was as long as it had always been. But there was also a shift.

More emphasis was being placed on technology than ever before, and the cost of maintaining that technological edge in the region in terms of equipment purchases and modernisation was more to the forefront of Defence budgets than ever before. It was to be a theme that would accelerate in the years ahead. This held more challenges for the leadership of the DRA as it continued to strive to retain the traditional ethos and traditions of the Reserve while adapting its understanding and knowledge of the ever-evolving Defence requirements for the Reserve.

There was no AGM conducted in 1998, presumably because of the distraction caused by the major organisation tasks around Reserve Forces Day that year. This caused some issues because clearly some branches had not paid the levy that year. The Association treasurer asked branches for their opinion of that. Was the levy on State branches an annual amount regardless or tied to delegates attending AGMs? In fact, it was an annual amount decided at each AGM for the current year. In this case, a provisional amount for 1998 had been decided at the 1997 AGM.

Although it seems remarkable that the DRA treasurer had to ask the question, it was more likely a reflection of the time-poor executive (and branch committees) of the Association which sometimes could temporarily lose track of the day-to-day work. It was a common theme running through the Association archives and records from the start of the Association. The DRA, however, like other associations run by volunteers with long-standing service in the committees, could rely on its institutional memory in that regard to solve short term issues like this one, and so it was in this case.[449]

In January 1999, the DRA president wrote to state branches to suggest a practical restructuring of the support mechanism for the highly successful Reserve Forces Day celebrations. In particular and in recognition of both the additional work load this had placed on DRA state committees as well as the need to broaden the appeal and standing of the RFD event, Major General Glenny suggested that separate committees be formed on which DRA could be represented but not lead (all states other than NSW had based their RFD committee around the DRA state committee). Glenny also saw that the DRA's role was being confused in some quarters as an event organiser rather than for its role as advocate for and representational body for Reserve issues.[450]

Meanwhile an article subtitled: 'The defence force is fighting a new battle—against funding cuts, economic rationalisation and low morale', in the *Sydney Morning Herald* in November 1998, focused on retention and change in the Services as a result of the 1997 Defence Efficiency Review.[451] The Reserves as always, also stood in the firing line of the Howard government's DRP, but this was by no means a 'new battle' for the Reserves, which had been fighting variations of this theme since 1948. By January 1999, morale in the Reserves had fallen even further when the Government removed the provision from industrial awards which allowed employees who were in the Reserves from taking leave for training. Allied to this the requirement for new recruits to undergo a six-week recruit course full time and the effect was no surprise; the Army Reserve was already 4500 short of its targets for recruitment.[452]

Partly due to concern about the effectiveness of the DRA in the face of these seemingly never-ending challenges to the existence and conditions of the Reserves, March 1999 saw Brigadier Andrew ('Andy') J. McGalliard RFD ED, both DRA vice-president and Victorian president, write to Major General Glenny in what was a continued discussion between them about the future and direction of the Association.[453] McGalliard acknowledged that the Association's effectiveness on major and national matters depended wholly on the president's success in lobbying key politicians and ADF senior commanders while the branches acted as State-based 'ginger groups'. That also placed an unreasonable demand on the president's time to identify issues, plan a course of action and then 'largely by himself, to put a case to Government'. McGalliard noted that while branches like his in Victoria were alarmed over the current situation facing Reserves and want to take action to change that, the 'ideas of what should be done are highly divergent.'[454]

McGalliard suggested to bring more people into the provision of management services to the Association, through the committees and using him, McGalliard, as a 'kind of operations manager' to coordinate issues which should be assigned to the appropriate State committee. This would take the weight off Glenny's shoulders to the extent that it should allow him to focus almost exclusively on the lobbying role and provide him with 'more ammunition'.

McGalliard, noting that the NSW branch would have extra demands placed on it by Reserve Forces Day (which by inference would also affect Glenny's time), identified the priority functions as:

- Defence policy.
- Role and function of Reserve Forces.
- Conditions of service.
- Welfare.
- Support for Reserve Forces. and
- Reserve Forces Day.

Several days later, Glenny reported to the State presidents and others of the outcome of his discussions with the Minister for Defence John C. Moore, MHR (Liberal/Ryan) and briefly with the Minister Assisting MINDEF and responsible for Reserves, Bruce C. Scott MHR (Nationals/Maranoa) on 24 March 1999. He reported that in the time available:

> ...it was impossible to outline our concerns and also counter daily contact he has with Defence and his own advisors. The Association is going to have to seriously review how we articulate our case. I do not believe our view and that of the Government are the same when we talk of the Reserve Forces.

The Minister told Glenny that he had asked Defence to conduct an internal review regarding Common Induction Training, but Glenny responded by saying that the Reserve was 'reviewed out' and that each review was an opportunity to attack the Reserve.... 'neither Defence nor the Government had any credibility left regarding reviews.'[455]

The Minister was not familiar with the Protection of Reserve Civil Interests on Callout—an issue about which the DRA had been lobbying on for years despite Lieutenant General Sanderson's sidelining of the issue in the past—nor was he across the Award and Individual Enterprise Agreements affecting Reserves. He wasn't aware of, specifically, that Reserve Leave was no longer an allowed condition, although the Minister did acknowledge that 'the whole industrial and employer scene needed changing if the reservists were going to be able to meet the demands being placed on them'. Other matters Glenny raised were noted, but no concessions were made. Although an invitation had been issued to the Minister to attend their annual conference, he did not seem aware of it nor 'that it was high on his agenda.'[456] Glenny drew no conclusions in his report, but it seems clear that he was tired and discouraged. The Defence report to the Minister did recognise the failings especially affecting recruitment and retention and looked for solutions, according to press reports at the time.[457]

Late in March, the Minister assisting MINDEF, Bruce Scott spoke to RUSI in Canberra. When discussing Reserves, which fell under his portfolio, Scott noted the 'discussion paper' being staffed in Defence to 'look at the very foundation of how we employ and support the Reserve component'. He said: 'We will see more integrated units, more High Readiness Reserve units and formations, and a greater reliance on the Reserve performing more specialist tasks'. Scott recognised the inflexibility of the existing Call Out Legislation and the need to address the weaknesses of the employer-Reservist relationship, not least the issue of leave for training. He also noted the need for job protection.[458]

The Government was finally responding to a barrage of criticism, not just from the Opposition in parliament, but also due to sustained questioning of policy by such organisations as the DRA and supporters of the Reserve across the country.[459] Continuing the pressure on Government, and in a direct approach to the PM with the blessing of the DRA president, the Victorian branch president Brigadier McGalliard also wrote to PM Howard on the future of the ARes in May 1999.[460] Among other matters, the letter expressed the view 'that what is required is not more reviews but executive action by Government to correct past errors.'

The letter noted that there had been a number of reviews going back over many years, with 'one common feature being that the long-term consequence of each has been a loss of numbers and capabilities within the ARes/CMF. The Reserve has now reached the point', it continued, 'that another such outcome would almost certainly encompass the loss of any effective Reserve Force from the ADF'.[461] In his letter, McGalliard raised a series of principles with the PM regarding Reserve service and contributions, the unique nature of Reserve requirements and of the flexibility to make reservists effective with in a 'one task, one service but many different contributions' approach. 'The fostering of a 'one-force' culture (as in the One Army campaign) ... must inevitably prove ineffective, he said, 'because it fails to recognise the realities of Regular and Reserve service...' Copies of the letter were also sent to the Minister for Defence and Chief of Army.

Meanwhile the Victorian branch noted in its annual report that there was 'continuing concern'—presumably by Navy reservists as well as others—over 'the use of the Navy's use of Reserve service virtually as the preserve of recently retired Regulars who can provide long periods of service. In addition to the loss of all 'off-the-street' reservists it also means that the Navy is feeding off itself in a period of reduction of numbers and is making no provision for any need for expansion in the future'.[462] Now six years on from the change to a tri-service approach, it seems that there remained little progress towards a genuine tri-service approach to Reserve matters within the DRA, at least as reflected in the branch reports.[463],

While the Navy and Air Force Reserve components were small by comparison to Army, those components remained well under-represented on their Association committees whether at executive or branch levels. NSW with its vice-president Squadron Leader Gilbertson, Victoria with vice-president Commander Hedges, and Queensland with vice-president Commander Hume, had at least started in the right direction. By September 2000, the NSW branch was reporting 379 members, with 292 ordinary and 87 life members, a far cry from the late 80s when the branch was struggling to survive.[464] However, the figures of only nine RAAF members and 16 RAN members showed there was still a long way to go in becoming a true tri-service organisation. As well, only 15 female members were listed among the membership numbers.[465]

The DRA Victoria newsletter of September 2000 had some trenchant criticism of the Navy's new pivot on Reserves. 'With the change from local Port Administration to PNF management, wherever that happens to be, Reserves have lost their focus and are no more than individual plugs for a depleted PNF structure', the report said, '...any career structure... is now non-existent...the focus for reserves has all but vanished.'[466] The PNF 'continually uses the argument that only PNF personnel are competent on modern Navy configurations', and, while the employment of reserves has been influenced by the adoption of integration, [Navy] 'has failed to define a clear role for reservists'.

With Navy removing two training ships and crews from use, 'the money effort and will [to build a credible Reserve] is not apparent today'.[467] The fact that Navy's *Reserve News* was to be subsumed into *Navy News* only added to the angst. RANR historian, Commander John M Wilkins OAM RFD, later noted:

> The new policies of 'privatising', 'outsourcing', 'selling' and 'integration' now sees the RANR role as a temporary staff replacement agency filling operational fleet vacancies arising from Defence inability to meet varying RAN permanent service recruiting objectives.[468]

The Victorian President's Report of September 2000 to the national AGM in September noted that 'the past year had seen a series of events having major impacts on the principal areas of interest of the Association'. These events included the ongoing operations in East Timor, the Government sponsored public debate on defence policy and the forthcoming White Paper along with the parliamentary inquiry into the Australian Army, and the proposed changes announced into Callout and Employer Support for the Reserves.[469] In the background, about 2000 reservists were deployed to support the Olympics in Sydney in September and October 2000.

The Parliamentary Joint Standing Committee on Foreign Affairs, Defence and Trade (JSCFADT) was tasked by the Minister for Defence to conduct an 'Inquiry into the Suitability of the Australian Army for Peacetime, Peacekeeping and War' with submissions called for by 2 July 1999. The DRA only had a few weeks to prepare a submission, but the deadline was driven by the reconstitution of the Committee when new Senators took their seats in Parliament from 1 July.

Presumably McGalliard's earlier suggestion to Glenny that he act as a coordinator had been accepted, for soon after, McGalliard set out the 'formidable list' of submission criteria which the DRA needed to address. It suggested that branch submissions be collated by the National Executive after 24 June and prepare the DRA submission. In the meantime, the DRA sought to make verbal submissions to the JSCFADT because, as McGalliard put it: 'This must increase our chances of having our ideas accepted'.[470] He gave no reason as to why he thought this would be the case. The Committee's report, 'From Phantom to Force' was completed in November.

The Australian Reservist was published twice in 2000, in February and again in November. By now calling itself 'The Official Journal of the [DRA]', it followed the style of the first edition but, with the third edition, started to develop a more organised format and presentation. The November issue provided a president's message from Major General Glenny, in which he concluded: 'This year will be the one that has the potential for setting directions for a Reserve to occupy its rightful place alongside the Regular component, where both components recognise the unique skills, characteristics and capacities they contribute to the Total Force'.[471]

Notably the issue also paid a tribute to Major General Cullen (who was to pass away in 2007, aged 98), noting his military contribution spanning 73 years and his ongoing and outstanding contribution as foundation president of the Association. Several shorter articles made their mark. Colonel Sandow in typically direct fashion started his by stating: 'There is a risk that our military planners will continue along the path of making small ideas look grand and making basic logic look small'.[472] The issue also reproduced the address at the national conference in July 2000 at Randwick Barracks Sydney by Laurie D. T. Ferguson, MHR (Labor/Reid), Shadow Minister for Defence—hence formalising the start of a tradition by the Association to invite both sides of politics to its conferences. By this third edition, *The Australian Reservist* had doubled in size and was to go from strength to strength.

Finally, after many, many years of representations, lobbying campaigns, and submissions of all kinds by the Association, its branches, and individual members, on 9 November 2000 the Government introduced the Defence Legislation Amendment (Enhancement of the Reserves and Modernisation) Bill 2000, with the primary aims to:

- extend the circumstances and the times in which the Reserves may be called out.
- repeal the *Defence Re-establishment Act 1965* which contains the current legislative framework for protection of the civilian employment conditions for reservists.
- facilitate the payment of financial incentives and compensatory payments to employers of reservists and self-employed reservists.[473]
- The tipping point for the Government was the need to 'top-up' a stretched Regular Army deployment to East Timor in 1999 and the need to call-up reservists for full time service including overseas (a Reserve rifle company also deployed to Butterworth in Malaysia, replacing long term rotations by Regular units). The long-held DRA position that the Reserve *was* needed was finally vindicated.

A Defence Consultative Committee (DCC, or the 'Defence White Paper Discussion Group') held extensive public meetings and private hearings; many submissions were received including from the DRA and senior DRA and other retired Reserve members. The committee comprised Andrew Peacock MHR (Liberal/Kooyong), former Liberal Senator Dr David MacGibbon, former Labor Senator Stephen Loosley and Major General Clunies-Ross. It was to report to the Minister for Defence by November 2000.

The process was met with a certain degree of scepticism by DRA branches. NSW branch noted in August: 'There was a general perception that the DCC is a flag waving exercise and that the major decisions had already been made'.[474] DRA Victoria president Brigadier McGalliard remarked:

> If this organisation has a purpose, it will have to justify it in the next six weeks. With the Peacock Committee commencing its research, all options concerning defence are open. What is acceptable to the government as reserve service, is not necessarily what the Reserve or the DRA would recognise as service.[475]

The Victorian branch meeting just before the end of 2000 expressed a range of concerns around the Community Consultation Report (CCT) by the Peacock Committee and the White Paper 'Defence 2000: Our Future Defence Force'. This despite MacGibbon, who claimed in the wake of the report to the *Australian Financial Review* on 10 November 2000, that Australia's military forces lacked the 'critical mass' necessary to maintain professional competence'. DRA members expressed the view that 'it looked like there would be no reform of the Defence Department' and that there was 'a very real concern that Reserves will be used as reinforcement and filler-uppers of positions, notwithstanding the sentiments expressed by politicians'. The minutes of the meeting recorded a new low: 'It is suspected that the real situation developing is that reservists will not get beyond the rank of Captain (Army). The capacity for the Reserve to shine is being eliminated—they use the neglect (of the Reserve) as an excuse for (further) neglect.'[476]

Despite the progress made, especially with the Call Out Legislation, it had been at a cost as demonstrated by the increasingly fatalistic, sometimes even cynical views, expressed at all levels as to whether anything had really changed in a positive way for the Reserves since the inception of the Association in 1970. Deep suspicion of politicians, defence bureaucrats and Regular Army hierarchy had become ingrained after years of seeing the Reserves bear the brunt of change to its detriment. Yet for all of that, reservists remained optimistic and remained determined to defend the Reserves, protect it, and enhance its place in the ADF. DRA leadership continued to support the Total Force concept, while pressing for positive change to support the Reserves.

As the DRA entered the 21st Century, it recognised that 'the issue of the reserve forces has been the subject of intractable debate.'[477] There had been many reorganisations and inquiries which impacted on the Reserves, the most prominent being, and not including many Army or Defence internals reviews, inquiries and reports along with numerous Parliamentary inquiries as well:

- Committee of Inquiry into the Citizen Military Forces 1972-74—The Millar Report.
- The Reformation of the 2nd Division 1981.
- The Reformation of the 3rd Division.
- The Sanderson-Nunn Inquiry 1987.
- The Force Structure Review 90/91.
- The JSCFADT Report—The Reserves, 1991.
- Disbandment of 3rd Division 1991.
- Project WELLESLEY—Rationalisation of Army Training.
- Establishment of the Ready Reserve Program, May 1991.
- Restructuring of the 2nd Division 1991 and 1994.
- Review of the Ready Reserve Scheme by John Coates and Hugh Smith, 1995.
- An Army for the Twenty First Century/Restructuring the Army, 1996-1999.
- The Defence Efficiency Review, 1997.
- The JSCFADT Report—From Phantom to Force, August 2000.[478]

However, as the JSCFADT noted in its report, 'despite all these reviews and inquiries, fundamental and sustainable reform to produce a useful Reserve has not eventuated'.[479] Colonel Maurie Ryan commented in his history of the MUR:

> The effect of many recent changes is to make the ARes accessible only to the chronically unemployed, the rootless seekers after jobs rather than careers, and ...students. The mature, stable and long-serving citizens who were previously the target of recruiting and retention efforts are now debarred from their right—if not obligation—to participate in the national defence.[480]

The challenges facing the DRA and its leadership had not gone away. The focus of the DRA had evolved with the evolution of the Defence and reserve paradigm over the many years since inception; its focus in the future would remain at heart that the Reserves continued to be 'of and from its community'. However, in the coming decade the Reserve would become highly valued as the ADF's operational tempo quickened and it came to rely on its Reserves like never before.

9.
Another New Era

Despite the excitement in September and October 2000 of the Sydney Olympics for which up to 2000 reservists had been used in a wide variety of roles around that major international event, the first year of the new millennium started slowly for the DRA. It faced yet another year of change and challenges as issues affecting the Reserves 'made haste slowly'. The end of 2001, however, would find the Reserves and the ADF in an entirely new paradigm. April saw Call Out Legislation finally in place which gave the government flexibility to call out the Reserves, for operations ranging from war to peacekeeping, to humanitarian assistance and disaster relief.

By late May 2001, the Government introduced an Employer Support Payment Scheme to compensate employers (and self-employed reservists) for allowing their employee reservists to undertake periods of continuous Defence service. Other new legislation offered various forms of protection for reservists, including protection against discrimination and employment, partnership, education, and financial liability protection.[481]

This was the *Defence Reserve Service (Protection) Act 2001*, which minimized some of the employment, educational, financial, and other disadvantages that Reserve members may experience when undertaking Reserve service. The Act also enhanced Defence capability by improving the availability of ADF Reserve members.

Following the hearings of the Peacock Committee and the release of the White Paper, 'which we believe contains both promises and threats for the future of the Reserves', the DRA continued to lobby the Government and the Defence Department. 'Bitter experience has shown that if we are not pushing forward, we are slipping back' warned the DRA Victoria president McGalliard.[482] Major General Glenny noted that that a submission to both

Ministers in April has contained 12 serious issues and although a response was received, 'it addresses none of the detail raised'.[483]

Reserve Forces Day on 1 July was larger than normal as it was the centenary of Federation, and 201 years since military forces joined the colony of NSW. The DRA, one of the sponsors of this 'National Reserve Forces Day', saw the event as a major opportunity to showcase to the Government and community the Reserve contribution. The DRA would march under its own banner.[484] However, in Victoria that branch had not marched on Anzac Day under its own banner as it wanted to make its first march 'a success', but did not do so again on 1 July, acknowledging that most members would march with their units or corps, while noting that 58 Navy reservists had marched, but no PNF.[485] Soon overshadowed by the DRA conference and AGM, held in Sydney at Randwick Barracks on 21 July 2001, the RFD march organisation and resources had once again stretched DRA committee and executive resources. While the success of the 2001 March kept internal criticism of heavy DRA involvement somewhat muted, the issue of DRA's commitment at the expense of other issues would return in years to come.

The conference keynote speaker was CA, Lieutenant General Peter J. Cosgrove, AC CVO MC. Cosgrove noted that the 2000 White Paper had 'changed the strategic role of the Army's Reserve component from one of primarily mobilisation and expansion to one that emphasizes its participation in the types of land operations in which our land forces may become engaged in the near term'. Cosgrove concluded by calling on the members of the DRA to join the fight to achieve the goals for the Reserve to ensure they were 'well-led, trained, equipped and employed...and valued by their colleagues in uniform'. In reporting Cosgrove's speech later, the editor of *The Australian Reservist* wryly commented: 'We have been fighting for these goals against the bureaucracy for decades and we are not about to give up now'.[486] Although reservists had been employed in East Timor, it had still been as individuals as much as sub-units, and essentially in a piece-meal manner.

Other presentations followed, including by Michael O'Connor of the ADA, on the changing nature of warfare; by CRES, Major General Garde with a detailed update on the state of the ARes; and by Major General Darryl C. L. Choy OBE RFD as Director General of the Australian Services Cadet Scheme. Choy noted that the Navy had 2693 cadets and Air Training Corps 6493 but Army had 16,571—25% of reservists were former cadets. Brigadier Paul Irving AM RFD, Deputy Commander Training Command and Major General Neil Wilson AM RFD, Commander 2 Division presented; they were followed by the Shadow Minister for Administrative Services, Laurie Ferguson, and Minister for

Operation TANAGER August 2001 in Maliana-East Timor. Private Greg Symonds, on patrol with No 4 Section, 11 Platoon, Delta Company, The Fourth Battalion Group. Greg is a Reservist on full time service whilst in East Timor.
(Defence Images)

Veterans' Affairs and also Minister Assisting the Minister for Defence, Bruce Scott, who gave the Opposition and Government perspective on Reserve issues.

At the AGM that followed, there was discussion about the conferences, gently prodded on by the national president. Should they even be continued? It was agreed that they should be. How were they best organised? It was agreed that the National Executive should run them, but the State branch involved needed to ensure there was an audience and also support conference logistics.[487] It was another small step in the evolution of the DRA's conference and its growing standing as an important annual event and the only Defence conference dealing with critical Reserve issues. Branch reports were received. The Tasmanian branch reported 122 life members and 153 ordinary members, of which half were serving members.[488]

The world changed on 11 September 2001 with the coordinated and strategic terror attacks on targets in the USA. Australia was quick to join the US in its newly declared war on terror and its defence posture rapidly changed. In his message in the November edition of *The Australian Reservist*, Major General Glenny noted that just six years before the Reserve could have provided formed units to protect vital national assets and infrastructure but which now had been 'forced to waste to insignificance and where its role is increasingly to fill out hollow Regular units.'[489] Returning to a recurring theme, Glenny reflected on the 'endless enquiries and reorganisations' and asked 'what reorganisation has enhanced rather than been another step to destroy the Reserve? The DRA', said Glenny, 'has always maintained that none of the supportive measures are introduced in advance of dramatic change and sometimes not until years after, if at all'.[490]

On 27 September, a speech was made to Parliament by Leo R. ('Roger') Spurway Price MHR (Labor/Chifley), Deputy Chair of the Defence Sub-Committee of the JSCFADT. His speech noted that the White Paper had committed the Army to provide a brigade on operations for extended periods and maintain a least a battalion group for deployment elsewhere, and he asked: can the Army do what it has been asked to do? Price particularly noted he was confident that after speaking to Reserve units and to branches of the DRA, 'they are prepared to embrace reform' to fulfill the roles assigned. Price said that the 'need to take difficult decisions to reform our Army is becoming more acute.'[491]

In November 2001, the Victorian branch president's annual report gave some interesting insights. It stated that 'with the turmoil in Defence, the situation for Reserves has notably improved, but their numbers are few'. The report stated that the branch needed more submissions and comments 'due to increased openness in the Defence debate (we recognise the need for more depth in some of our policies), as the Victorian branch seems to be the ideas factory for the National Association'. NSW branch may have contested that assertion, but NSW was certainly not as active as it had been in the past in the pursuit of DRA objectives.[492]

The Victorian report also noted that while the DRA at the national level seemed to have good contact with Government, locally contacts seemed to mainly with the Opposition. Reserve Forces Day had been on the annual calendar now for four years supported by the DRA, but the report noted that in future, the ADF would be managing it. On the one hand this removed some of the financial and resource pressures felt by the DRA and other organisations but on the other, there was some feeling of being sidelined from its own parade.[493]

The 11 September 2001 terror attacks in the USA had already dramatically changed the security and defence dimensions for allied countries, including Australia. Operations in Afghanistan and in the region led directly to a change in attitudes to the Reserves, already influenced by the good performance of Reserves in East Timor and later Bougainville.[494] While this would take some time to evolve, and would be accelerated under the returned Coalition government, other issues of concern to the DRA continued to move along, sometimes at what appeared to some observers at least, to be at a glacial pace. One major change that did come into being in December 2001, however, was the change in title of the 'Inactive Reserve' to the 'Standby Reserve' under new legislation which also enacted the provision for enforceable call-out of those who transferred into the new component.

The role of the DRA came into focus as 2002 progressed. While the cooperation with Kindred Organisations continued, none of them would look after the Reserves as such so the DRA remained the main organisation doing just that. There remained plenty of issues to consider and try to influence—some of long standing, others new to the changed defence environment. The DRA remained engaged with service and welfare, the pay system (rates, taxation, superannuation, long service leave etc.), cadets, how the Reserves were used, types of Reserves, recruiting and retention, and general ongoing support for the ADF and the Reserves in particular. The DRA was closely following the Review of Defence Force Remuneration (the 'Nunn Report') which was described as a wide ranging, independent study, conducted at arm's length from the ADF. It had reported to Government in 2001 and the ADF response was awaited in March 2002.[495]

Problems with Reserve Forces Day in Victoria surfaced in a letter from the president of the Victorian branch of the DRA, Brigadier McGalliard, to Brigadier Peter D. Alkemade RFD ADC, commander of 4 Brigade, in August 2002. Noting that the march in Melbourne was smaller than previous years, McGalliard saw that more practical arrangements, nonetheless, had been put in place 'to ensure that the event is readily accepted by the authorities and satisfies both the commemorative and more pragmatic goals.'[496] McGalliard wrote:

> Some extra-curricular arrangements were inevitable in earlier years when it was uncertain whether the Day would continue and everything was on an ad-hoc basis but cannot be continued if the event is to be ongoing, even if not on the same scale or format. …whether or not [RFD] was a good idea when proposed in 1997, and there was considerable diversity of opinion about it, it is now so firmly established in many people's minds that to abandon it would be harmful to all involved, the ADF, and especially the Reserves, the Associations, the DRA etc.[497]

Certainly, the DRA Victoria branch experience had been woeful, describing it in its minutes in October that year as a 'total and absolute failure', with the Defence Reserves Support Committee commenting to the branch: 'The march was a disgrace, with no uniformed Army involvement, and estimated 1.000 involved'—5000 had been expected. Australia-wide publicity for the event had a budget of only $120,000—'a joke'.[498] The feedback resulted in a directive by the CDF, General Cosgrove, to Major General Garde as Head of Reserve Policy to coordinate all Defence Reserves programs, including ADF support for Reserve Forces Day activities, from then on.[499]

Meanwhile the DRA conference and AGM for 2002 was held over 24/25 August, again in Sydney. The opening address was given by the new Chief of Army, Lieutenant General Peter F. Leahy AM. Leahy stated that Reserves were now 'front and centre' within a joint war fighting, seamless ADF model when reservists would 'really achieve their full potential', among other things, to meet the 'significant round out and rotation support' for Army.[500] Presentations followed by familiar faces such as Brigadier Irving ('Training Initiatives for the ARes'), Michael O'Connor from the ADA ('A 25 year retrospective—Responding to Change'), Major General Garde as Head of Reserve Policy (HRP), and Major General Wilson as Commander 2 Division.

New faces included Dr Alan Ryan from the Land Warfare Strategy Centre, who pulled no punches in his talk 'Back to the Future: The One Army Concept in a Time of Change':

> The real issues in the Reserve debate are obscured by a century of myth making by an often overly romantic Reserve lobby and the equally destructive counter-cynicism of many Regular soldiers. The real issue is how in a changing society the ADF can maximise its access to, and use of, the limited human resources available to it.

Graham J. Edwards MP (Labor/Kalgoorlie), Labor's Veterans Affairs spokesperson—and the only veteran serving in that Parliament; Danna S. Vale MHR (Liberal/Hughes), Minister for Veterans' Affairs and Minister Assisting the Minister for Defence, gave Opposition and Government perspectives. Among other speakers, DGN Reserves, Commodore F. Karel de Laat RFD, reported on the current state of the Navy Reserves.

The AGM was well attended, except for the 'semi-official NT' branch, which had no delegates.[501] At the AGM, president Major General Glenny noted that the DRA 'structure is a concern, but funds available to national are an issue of greater concern'. He went on to say of Association membership, 'we are not doing well with uniformed personnel, which is the reason for our existence'. Perhaps in reflection of this, it was agreed by those at the AGM that 'poor attendance at this year's conference was a real concern'. NSW alone had sent out 390 invitations and the president a further 74.[502]

Glenny noted that the DRA now had representation on the RSL's national defence committee, and that there had been contact on a regular basis with both Opposition and Government including the PM, Defence Minister and others.

The meeting then discussed a range of matters including whether to update the Manifesto which had been prepared under Colonel Sandow's time as president (Sandow thought it was not worth the effort); Glenny said that the Association needed to be able to respond to issues more quickly. The project officer idea was discussed, and several suggestions were put forward, with once again conditions of service, tax and superannuation being the prominent issues for projects. The meeting also agreed to support the idea of a Defence Service Medal and would work to have Government accept the concept over the next 12 months, and that the Tasman Exchange Scheme should become a national scheme for the DRA, administered by the Tasmanian branch.[503]

With the 2003 conference planned for Melbourne, the Victorian Committee started work almost immediately by forming a sub-committee to develop not just the logistic plans but to improve on the 2002 conference by different stratagems such as ensuring a Minister was there, and having a focused programme around a theme, with speakers selected for their relevance to that theme (these ideas reached fruition at the 2004 conference). In another development for the branch, it had been agreed from the national AGM that instead of an update of the Manifesto, short policy papers would be developed instead with Victorian branch as national writers. As a start point, an update of the DRA submission to the Defence Review 2000 would be undertaken.[504]

Major General Glenny's message in *The Australian Reservist* in November 2002 followed soon after the October terror attack in Bali where 202 people were killed including 88 Australians. Glenny questioned the rationale behind the primary aim for Reserves of individual reinforcement and augmentation of under strength Regular units, claiming it was arguable and supportable only for peacekeeping/peace enforcement activities. When the Reserves undertook vital asset protection roles between 1989-1995 it had the units and numbers to do so. Now the Reserve was seriously understrength. Glenny saw the low numbers as 'another opportunity for detractors of the concept of Reserve forces to launch further ill-informed attacks'.[505]

At the Victorian branch AGM on 10 November 2002 the annual report noted a healthy membership of nearly 200 individuals plus memberships from affiliated bodies, such as regimental and corps associations and other service groups. The committee noted without fanfare that its first female reservist, Major Elizabeth D. Bedggood RFD had joined the committee. The Victorian branch committee also kept in closer touch with the National Executive through its secretary, Lieutenant Colonel G. J. Sharkey RFD, who lived in Melbourne and often attended meetings of the Victorian committee. Notably the report dwelled on the raising of a company of reservists for service in Timor, the first time a fully formed Reserve unit had been so raised—certainly a 'milestone event—but nonetheless a composite company given that no one Reserve unit could provide the full complement.[506]

In December 2002 Major General Glenny sent a message to Danna Vale (Liberal/Hughes), the new Minister Assisting the Minister for Defence, to raise the serious issue of Reserve recruiting and training. He also raised the cancellation of recruit training courses for reservists in December and January 2003, thereby preventing the very recruits at the end of school or university years who could possibly attend from joining the common recruit training (CRT).

Glenny noted that although recruited into the Reserves only to be prevented from training at Kapooka base in NSW, these recruits could not even be trained by their own Reserve units. This, he said, 'seriously challenges the Government and Defence commitment to rebuilding the Reserve', noting that the numbers of the Active Reserve had fallen to about 13,800 against 23,000 only six years ago.[507] This concern may have been regarded by some as unimportant in the wider scheme of things, but it must also be set in the context of growing belief that Australia could become embroiled in a general war in Iraq—and what role would reservists play if that came about?

The cancellation of recruit courses in December 2002 and January 2003 meant recruiting, said Glenny, 'had a hit a wall', with 600 reservists Australia-wide being prevented from attending the CRT when they were available and being panelled for March when they were not available. A paper written for the DRA by Colonel Rainer H. Mayer-Frisch MBE, 'A Strategic Plan for the Revitalisation of the Australian Army Reserve', had been circulating and attracted much interest by January 2003. Mayer-Frisch argued that retention rather than recruitment should be given priority as recruiting had not met its targets for the previous five years. It was subsequently circulated as a background paper and a personal view but not as a DRA paper.[508] The Victorian branch was in the meantime working on an update to the original branch submission to the Peacock Committee in 2000.

By July 2003, with the return of the first Reserve company to be deployed to East Timor, the implementation of the High Readiness Reserve component within the formation of the anti-terrorism force and the deployment of Reserves in Victoria as part of the response to summer bushfires, reservists were in the spotlight like never before. Already the pessimistic prognosis expressed in the September 2000 newsletter of the Victorian branch—that the Reserve had 'been reduced to a component regarded as a resource of doubtful value...and destined for obscurity'—was being overturned by the dramatic changes in Defence posture caused by international events.[509] It could be said that if it was not for that, the downward trajectory of the Reserves 'to obscurity' was indeed a foregone conclusion.

The DRA conference and AGM was held in Victoria at Simpson Barracks, Watsonia over 30/31 August 2003. Perhaps because it had not been held in Victoria for some years it had a larger than usual attendance, although credit for the attendance was due as well to the planning put in by the Victorian branch. Major General Glenny opened the conference emphasizing a range of issues facing the Reserves such as critical shortages in key personnel, ongoing recruiting shortfalls and uncertain career paths. Major General Francis ('Frank') X. Roberts AM, Deputy Chief of Army, stressed the need for the revitalisation of the Reserves and raised the question—'A Reserve Army or a Reserve for the Army?'

On that topic Colonel Mayer-Frisch spoke to his paper on 'Revitalisation of the ARes'. Major General Cooke—the DRA representative on the Defence Reserves Support Council—reported at length on that organisation. Cooke's presentation on the DRSC provided an overview of the relationship with the DRA, noting that the DRSC had come a long way since the first committee was formed in the 1970s as the CESRF and how in the period 1985-1988, there was only one dedicated Major who had responsibility for the scheme Australia wide. DRSC now had a budget of $740,000 and was fully integrated within Defence and Government.[510] The presentation by ACDF-R and Head of Reserve Policy (HRP), Major General Garde, included an overview on recruiting and the capacity for Reserve surge capacity and sustainment.[511]

Dr Ryan from the Land Warfare Centre 'referred to controversial aspects'—such as the lack of contemporary combat experience of Reserves, that there was no need for a brigade or divisional structure in the Reserve and that Reserve officers would not ever be in senior command above sub-unit level. There were reports on the Navy Reserve by Commodore de Laat, updates on Reserve Branch at AHQ from Brigadier Neil Turner AM RFD, ADC DGR-Army; and a report on RAAF Reserves by Air Commodore David J. Dunlop CSC, DGR-AF.

Brigadier Wayne L. Dunbar CSC RFD, of 4 Brigade reported on developments in 2 Division, followed by Brigadier Ian B. Flawith CSC who presented on Civil Military Cooperation (CIMIC). In the following Q & A, the question was asked:

> Following the commitment to East Timor, new Reserve capabilities were promised including port handling, water transport, interpreters, specialist IT, laundry and CIMIC. Of these only one was raised, it is understrength and not used in the Solomon Islands. What happened?[512]

Finally came a provocative presentation (that is, to some of those attending), 'The Death of Military Thinking' by former Lieutenant Colonel, Neil James, the new executive director of the ADA, described as 'an independent thinker'. James made several fearless points, such as the Australian military was under civilian bureaucratic control and that this was dangerous, enabled in part because the organisation was too large for just one minister.[513]

The AGM included a longer report than usual from Queensland branch. In its annual report for 2002-2003, it detailed how it had 'endeavoured to take a new approach to the way [it] conducts its business', by, for the first time, developing a strategic plan which set out seven goals against which it could measure its performance. This had resulted in a higher level of participation by its members (54 financial from a list of 90) over the past year. A highlight too was progress towards full tri-service representation with each service now represented on committee.

Queensland branch president Brigadier Peter Rule AM RFD participated in media activities around the issue of Ready Response Forces within the HRR and the Reserves' ability to deliver capability in this regard.[514] Other branches also reported strong membership numbers, but SA continued to struggle with just 29. It stated that as available income was all taken up with its own administrative and running costs, it did not think it could pay the annual levy nor afford to send delegates to the next conference and AGM.[515]

In a paper prepared for a meeting with the Victorian Liberal Party Defence Reserves Sub-Committee on 15 September 2003, Brigadier McGalliard stated, in relation to tri-service Reserves, that 'The perception of Army only remains hard to combat for many reasons. However, there is great potential to significantly increase the Reserve role in RAAF and RAN as well as widening its contribution to the Army'. He went on to say: 'The failure of the RAAF and RAN to make meaningful use of the Reserves is a long-term loss which continues. There are indications the RAAF is beginning to move but not the Navy to date'. He did not elaborate.[516]

Major General Glenny once again put in a strong message in the November 2003 edition of *The Australian Reservist*. He strongly criticised the Common Induction Training/Common Recruit Training courses, noting that they had been opposed by the DRA for the past eight years and stating that the concept of One Army had in effect been used as a cover for the introduction of courses like CIT/CRT which were too long for any employed Reservist

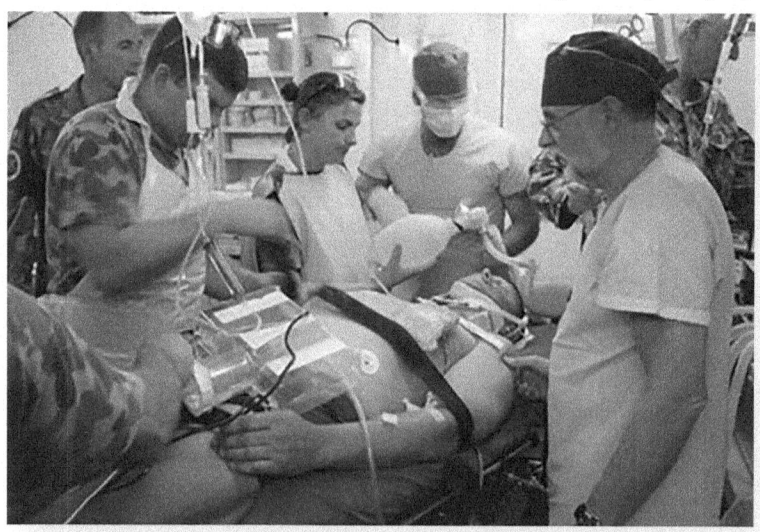

Commander Hamish Foster (right), chief surgeon of the new First Aid Health Facility at Camp RAMSI, prepares for an operation during a simulated casualty exercise during Operation ANODE, Australia's military contribution to the Regional Assistance Mission to the Solomon Islands (RAMSI). Commander Foster, a Townsville-based RAN reservist, also served in Vietnam and in the Gulf War. (Australian War Memorial)

to attend. The length of all employment, trade and promotion courses had now become a threat to the viability of the Reserves and especially its officers, warrant officers and NCOs, he said. Glenny also criticised the mandatory two year posting demanded of both Regular and Reserve officers, which caused 'unnecessary turbulence, loss of community contact, unit corporate memory and traditions'. In addition, Glenny decried the lack of recognition and unacknowledged contribution of Reserves within the 'One Defence Force' concept, even though half of the Defence Force were reservists.[517]

In early 2004 Glenny continued his round of 'DRA diplomacy', engaging with a range of Government and ADF figures. One topic under discussion—the 'latest' Military Compensation Bill was before Parliament. He also attended a briefing by the CDF of senior retired Reserve officers in Sydney in March and a briefing by Land Commander Major General Kenneth ('Ken') J. Gillespie AO DSC CSC, also in Sydney, in May 2004.[518] DRA input in 2003 into the Ken Aldred-led Liberal Committee report 'Revitalising the Reserves', presented to the Minister for Defence and the PM just before April 2004, was reflected both in the report and a hard-hitting article in *The Australian*—'Defence has Little in Reserve.'[519]

Meanwhile Major General Cullen met with the CDF and Minister for Employment Services Malcolm ('Mal') T. Brough MHR (Liberal/Longman), a former Army officer, on 11 May 2004 to provide a 'wish-list' for the Reserves.[520] The outcome is not known, but it was remarkable that Cullen was still advocating on behalf of the Reserves, and still able to gain an audience with the CDF and a senior minister. Major General Glenny later acknowledged 'the unfailing support and advice provided by Major General Cullen...never afraid of a fight, always happy to phone a key figure and always available.'[521]

The August 2004 DRA Conference was once again held at Watsonia in Melbourne, the organisers emphasising the change from 'reporting on the last 12 months' events to the future and how the Reserves will evolve; interact rather than react.'[522] The theme of this conference was 'Defence Reserves: Future Directions! Where to and How?', with the theme indicative of ongoing DRA concerns for the future of the Reserves at a time when significant changes were being proposed. Major General Glenny opened the conference followed by Major General Cooke who presented a paper on 'What the Reserves Should be, what is preventing this?', 'expressing the opinion that the 'One Army' concept had been "over-implemented' and that 'Total Force' is a more appropriate concept'.[523] Cooke noted that clear roles and tasks aided retention, and that the 'Ready and Relevant' Reserves concept had to recognise the difference between FTD and PTD.

Frances ('Fran') E. Bailey MHR (Liberal/McEwen) Minister for Employment Relations and Minister Assisting the Minister for Defence, spoke to 'Australia's Defence and the Reserves—Present and Future'.[524] The Victorian branch conference summary especially noted her statement that recruiting functions would be returned to units. Others found her prepared remarks 'a disappointment' but also that her additional comments were 'insightful'.[525] Lieutenant General Leahy spoke to 'The Army and its Reserve—Present Problems, Future Opportunities'. Later, he would publish many of these views which resulted in a DRA response as will be seen. Noting that change was accelerating he also stated that brigade and divisional structures would remain. Was that a direct response to Dr Ryan's paper the previous year? Leahy's comments implies there had been a backlash of opinion against Ryan's assertion that these traditional structures needed to go.

Roger Price, MHR, giving the Labor Party view of the Reserves, reminded the audience of the diminishing defence experience among current politicians and lack of addresses on defence topics in Parliament.[526] Major General Peter F. Haddad AO, Commander Joint Logistics, spoke to the topic of 'Reserve involvement in Tri-Logistics—Now and Future', reviewing the complexity of joint service logistics and the opportunities for reservists in the Logistic Support Force. Major General Richard G. Wilson AM, ACDF-Reserves/HRP, detailed Reserve numbers (under 2000 in Navy, over 2000 in Air Force and about 16,000 in Army) noting that lack of superannuation was a significant deterrent to retention. Wing Commander Richard J. Bluck AM RFD, Chairman of the DRSC, provided a report on that organisation and the results of surveys noting that reservists wanted access to service medical/dental and superannuation and opining that unless serving reservists concerns are addressed, recruiting efforts are largely wasted.[527]

Ken Aldred, former MP and Reservist, who had been very active on behalf of the Reserves in Parliament in years gone by, remained engaged with the Reserves through his leadership of the Liberal Party's Defence Reserves Sub-Committee. He spoke to the Sub-Committee's paper presented to the Government in April—'Revitalising the Reserve'. In view of the growing use of Reserve Commandos, which would expand dramatically with the commitment to Afghanistan, Lieutenant Colonel Anthony John CSC, CO 1 Commando Regiment spoke to 'New Operations—Reserves in Special Force Operations'.[528] The conference also saw two open forums, 'Total Force Concepts' (adjudicated by Neil James from ADA) and 'Where to Now?' (adjudicated by Major General Michael ('Mike') O'Brien CSC). Overall, about 100 attended. In closing the conference, Major General Glenny emphasised that the DRA 'must remain dogged and determined in the pursuit of solutions'.[529]

For the AGM which followed, Glenny made a lengthy report reflecting on the challenges for Defence reservists and the DRA response and action plan. The changing roles and tasks of the Reserve and the increasing challenges of excessive training to ensure that reservists are 'relevant and ready' were identified as the new challenges for the traditional Reserve. As Glenny expressed it, the specific challenges and turbulence of these issues would be compounded by Defence's inability to recruit and retain personnel. Within the DRA itself Glenny saw the challenge of recruitment for membership as well, as reflected in branch reports. He noted that the DRA had been 'singularly unsuccessful' in recruiting all the two-star officers (with the exception of Major General Garde), since Glenny had retired in 1994—'numbers are critical in order that we are not seen as a splinter group of the old and bold. The input of serving members ensures that we represent current issues.'[530]

Unusually, Glenny called for a revisit of the Constitution to redefine the roles and authority of the national president, executive and State branches, including branch direction and activities and the 'safeguarding of the integrity and standing of the DRA'. While providing no details behind this statement, it may well have been a reaction to internal criticism of some State branches which had started to be less effective as DRA lobbyists. Some critics highlighted what they regarded as 'busy work' not critical to pursuing DRA objectives. Examples cited were activities such as school history class programs, military history tours abroad, attending Defence related events and functions unrelated directly to the work of the DRA, or for some, over involvement in the RFD marches or becoming involved in kindred organisations. Those relationships are important, said Glenny, 'however the overriding objective must be the betterment of the DRA'. Yet, it was equally critical to the DRA that the organisation marched forward together, and Glenny, as its long-standing president, knew that any review of the Constitution had to be put forward carefully.[531]

Glenny also noted that it was time the DRA moved to develop a website. '...we are light years behind most other organisations', he said, 'and I am often asked by media what [website] they can look up'. He also questioned what use State branches were putting to editions of *The Australian Reservist* (there was no edition published in 2004). It came back to recruiting for the DRA itself. Glenny lauded the way the National Servicemen's Association had managed to grow from a limited base—he hoped that as it widened its base it did not draw reservists into its fold as well. Last Glenny called for a review of the national conferences. While the aims of the conferences were excellent, and presenters of high quality, how was it that they could not develop the same numbers of 150-200 as other defence conferences, and, importantly, why were they held at a loss, he asked? Glenny concluded his report by saying: 'We need to end the AGM with a clear set of objectives and the intention to achieve them.' [532]

Shortly after the Federal election in October 2004, which returned the Howard Government, Lieutenant General Leahy, Chief of Army, published an article in the *Australian Army Journal* 'The Australian Army Reserve—Relevant and Ready', in which he emphasised the vital contribution of Reserve capability to the ADF. [533] Certainly, Leahy's article was criticised. To some it reflected an attitude of mind in which the Reserve was still seen as separate from the Army itself. Even though recruiting was supposed to revert to units, the policy was four years old but there had been no action on it whatsoever. [534] The DRA saw this as an opportunity to respond with its own views.

In a comment on the draft response to be published in *Defender*, Colonel Sandow, the former AResA president, noted: '...it is better for us to "stick our neck out" rather than be too courteous. Only in this way can we excite the kind of reaction we need to stimulate argument and (thereby) extend the period over which the debate is kept running...Better to get a few black eyes now and keep the tempers high!'.[535] Yet the DRA itself was also seeking to get Reservist contributions recognised as such, for example involvement in the tsunami relief efforts in Indonesia in late December 2004. [536]

Subsequently the DRA was able to publish a critical response to Leahy's paper in the ADA's journal, *Defender*, in early 2005, with the help of ADA executive director Neil James. The article, 'Matching Performance to Promise—Rebuilding the Army Reserve', by DRA president Glenny, was comprehensive, direct and with a simple message: 'To enable the Army Reserve to deliver the capabilities implicit in its task list it must be nurtured, encouraged and supported with earnest, non-partisan and genuinely purposeful attention.' [537] Glenny called for 'urgent action' on several steps:

- Army Reserve training to be unit-based and not contracted out.
- Prospective ARes recruits complete their training over a period of two years through a mix of centrally and regionally conducted modules that best suit their civil employment circumstances, personal education program and availability throughout the year.
- The ARes be formally and directly represented at all levels of equipment acquisition so capabilities mainly or partially supplied by Reservist members and units are not diminished by ill-informed or poorly considered decisions.
- Reserve budget components be quarantined.
- Conditions of service for Reservist personnel be completely overhauled...
- The capabilities to be maintained by Reservist personnel promised over recent years be raised and equipped to full strength. [538]

Meanwhile, in October, during the 2004 election campaign, PM John Howard promised a comprehensive review of Reserve remuneration arrangements. By November 2004, a Conditions of Service paper had been circulated from Victoria to branches. By December the DRA had formulated a clear policy on conditions of service for reservists, noting that the most recent Review of Defence Force Remuneration (the Nunn Report) was conducted in 2001 and that most of the recommendations for enhancements to reserve conditions of service were not implemented by the Government.

In short, the DRA believed that Reserve pay rates should be fair and rational, and that conditions of service have an important effect on recruiting and retention. And presciently, in light of the many subsequent changes coming to the Reserve in 2006, the DRA believed that Reserve Conditions of Service should support capability.[539] The question of taxation of pay for reservists was the next issue for discussion. In the meantime, two other policy papers on Reserve roles and functions and Force Structure, both prepared by the Victorian branch, were also in circulation.

Despite Howard's announcement of a 'comprehensive review', in fact, as the DRA saw it, 'the influential hierarchy of the Department of Defence seem determined to frustrate our elected government by side-tracking reform attempts particularly in the area of service conditions.'[540] The Reserve Remuneration Review got underway in February 2005; in May the CDF set out the Terms of Reference to the Head of Reserve Policy (HRP) and Head Defence Personnel Executive.[541] It was however, no longer a 'comprehensive review', but an internal working party.

In July 2005, Major General Garde was asked to represent the DRA at the 'inquiry', but with the short time frames there was no time for this to occur. At best, Garde was able to brief Captain Peter J. Quirke RAN who was chairman of what had become the narrowly focused internal working group in August, providing 10 points from the DRA perspective for consideration by the ADF and Services. However, the Review reported in September 2005 and to the Minister in October—but had still not been released by the time of the DRA's August 2006 AGM.[542] By that time, Garde reported that progress had been made in only three of the ten areas noted to Quirke a year before.

Thinking 'outside of the box' and in an effort to engage with every possible means to influence the Government around the issues, Garde even arranged for a dinner on 3 April 2006 with the then Secretary of the Australian Workers Union, William ('Bill') R. Shorten, along with Major General Barry, Brigadier McGalliard, and Lieutenant Colonel George. Shorten found the discussions regarding poor recognition of the role of Defence Reserves 'most interesting and thought provoking'.[543] Eighteen months after the review was announced, 'no outcomes of the review have yet been made public, there has been no change to reservists' conditions of service and Defence Reserves strengths and capabilities continue to decline.'[544]

Meanwhile the Victorian committee continued its development of short DRA policy papers, approaching final drafts of the Conditions of Service policy and with new papers on Recruiting (tri-service) and Common Induction Training to follow (the CIT document was sent to the national president in April). Concern was later expressed that although the Recruiting paper was supposed to have a tri-service focus, it remained concentrated on Army.[545] This highlighted some of the challenges of the tri-service approach.

For example, most Navy reservists were ex-PNF. Navy reservists were not in formed units like those in the Army and the most effective method of recruiting was no longer available with the closure of the Naval Reserve base HMAS *Lonsdale* in 1992; recruiting since then only recruiting Navy Chaplains, lawyers, and doctors.[546] Developments also continued on the DRA website. Like many of the issues tackled by the Association over its many years of service, creating and managing a website was just another issue to find a solution for. In this case however, not so many of the members were experienced in doing so with confidence and eventually an outside 'consultant' was brought in to manage the site.[547]

The theme at the DRA's 2005 conference and AGM, held over 27/28 August at Watsonia, was 'Rebuilding the ADF Reserve Forces'. Major General Glenny, in his last conference as DRA president, opened proceedings. Following a presentation by ACDF-R/HRP Major General R. Wilson, the keynote speaker took the stand. Major General Jim Barry would be formally elected as incoming DRA president at the AGM next day. He noted that following the 2000 Defence White Paper, 'despite the best intentions, neither the strength or equipment procurement nor expenditure of the [2000] budget had been achieved'.

He said that recruiting targets had not been met for the past two years and stated that 'in fact, the Reserves are at their lowest ebb in their military history…The goal of the conference then was to identify the problem areas, decide on the best way forward and rebuild to provide an effective capability for the ADF.'[548] Barry also reported that Active Reserve numbers had fallen to between 19,000 and 21,000 with 2500 in the Standby Reserve, 'but Defence reportedly does not know their details.'[549]

A range of speakers followed, but the DRA scored a coup with the participation of Air Chief Marshal A. G. ('Angus') Houston AO AFC, the new CDF who had been appointed in July. He presented a paper 'Optimising Reserve Capabilities'. Parliamentary Secretary to the Minister for Defence, Teresa Gambaro MHR (Liberal/Petrie);[550] Employer & Community

Lieutenant Mathew Birmingham, RANR, stands beside one of the main engines in Number One Engine room onboard HMAS *Manoora*. Otherwise a local lawyer, Birmingham joined HMAS *Manoora* in the lead up to the Commonwealth Games.
(Defence Images)

Relations Chair at DSRC, Leneen Forde; Shadow Minister for Defence Industry, Procurement and Personnel, Senator T. Mark ('Mark') Bishop (Labor/SA); and executive director of the ADA, Neil James all presented papers in turn.[551] Neil James became a regular speaker or adjudicator at DRA conferences.

The post conference synopsis was written by Major Ian S. Rainford OAM ED JP. He commented that Major General Wilson's address, for example, 'sounded a strong note of optimism', but:

> ...the concentration on the High Readiness Reserve (HRR) must lead to the total neglect of everything else which means the bulk of the ARes will die in five years or less....there is a strong feeling among two or three star Regulars in high places, that the time of the Reserves has passed and it is now high time for professionals only. It is going to be a hard fight and we must not underestimate the enemy.[552]

These comments were illuminating. It appears, on the face of it, that for some retired Reserve officers the long-standing suspicion of the agenda of the Regular hierarchy remained alive and well, but Rainford's commentary on the HRR was well founded. Fears of the demise of the Active Reserve, being stripped to man the HRR, could not be dispelled easily. At a time when there existed a 'chronic recruiting shortfall and woefully inadequate and wasteful retention rates', the worst fears of some for the Reserve seemed to be being borne out.[553]

The AGM for 2005 followed the national conference—there was a changing of the guard. Long serving DRA president Major General Glenny stepped down and was replaced by Major General Barry. Glenny had been president for nine years and in that time, it was recognised that the Association 'had moved from simply being a pressure group to a body that is invited to participate'. In another resolution the AGM recognised that during his presidency 'the DRA had defeated the Army21 push for the demolition of the Army Reserve and had received the support of the Government and the Opposition, CDF General Cosgrove and CDF Air Chief Marshal Houston for the enhancement and restoration of the Army Reserve'.

In accordance with the Association's practice, Glenny was awarded Life Membership and nominated to replace Major General Cooke as the DRA representative on the DRSC national council.[554] Of course, there were major changes in the strategic outlook in this period as well—as the ADA's Defence brief in late 2006 put it:

> Strategic events and their resultant lessons since the 1999 East Timor intervention, and the 9/11 and Bali attacks, have well and truly driven a stake through the heart of the disproven "defence of Australia" dogmas of the 1980s and 1990s.[555]

This outcome had eventually worked in Major General Glenny's favour given that he had not deviated once in his advocacy for a strong Reserve and the need for an enhanced role. His position also gave the DRA enhanced standing as well, and his views were strongly justified as time went by. Attended by 125 present and with the new CDF speaking, the conference had been a 'roaring success'. The Victorian President, Brigadier McGalliard, noted that 'now the DRA get *invited* to talk over issues rather than having to *push*....'[556] Major General Barry immediately formed a 'task force' of himself, Major General Garde, Lieutenant Colonel George, and Brigadier McGalliard to brainstorm ideas, consider options and pathways and best responses to issues as they arose, whether administrative or policy driven.

Meanwhile the branches were quietly working away at representing the DRA in a range of functions and activities in their own States as resources and the level of Reserve presence in the State allowed. While they were not always at the forefront of the major policy papers and submissions their efforts all contributed to the sum of the parts that was the DRA and its national profile, including with media. Everyone agreed that there was never enough press coverage of the Reserves—but that had been a catch cry for the entire period of the Association's existence. However, now the National Executive was also trying to get on the front foot with media releases responding to Government announcements and following DRA conferences. Some success with this new strategy was seen in February 2005 when an article in the *Australian Financial Review* resulted from a DRA Media Release 'Improved Conditions of Service crucial for Defence Reserves Strengths'.

As always, it was difficult to get press coverage of Reserve issues, but branches did what they could as well. For example, in October 2005 Lieutenant Colonel Alexander, DRA South Australian president, was able to have published his response to an earlier letter to the editor mentioning Reserves.[557] In NSW the branch had an active press group identifying press errors in published articles and which prepared rapid responses. One of the decisions of the 2005 AGM had been to have one of the Victorian committee members, Commander Peter Hicks RFD RAN, take leadership in guiding the development of the website for the DRA. This at least was a start point to enter the internet world and seek an effective online presence. The State branches were, on the whole, slow to come on board, but that may well have been more about resources to administer State 'home pages' than a lack of interest.[558]

In another of the annual briefings by the CDF for senior retired Reserve officers, in late October Major General Barry attended the briefing by the now VCDF Lieutenant General Gillespie 'and afterwards had a positive discussion about the Association's role'.[559] Brigadier McGalliard, the Victorian branch president reported to the Victorian AGM on 17 November 2005. Among other matters in the report, it was noted that one outcome of reservists being increasingly involved in overseas deployments was a growing cooperation with other organisations [but] especially AVADSC, which involved every welfare group linked to ex-servicemen and women—'the welfare needs of reservists could potentially become enormous.'[560]

10.
Raise, Train and Sustain

In December 2005, the Government formally launched the Hardened and Networked Army (HNA) concept, under development for more than 12 months. It was designed to enable the Army to sustain a brigade on operations for extended periods and maintain at least one battalion group for deployment elsewhere. The Army Reserve was asked to provide key components to the HNA. The concept moved the Reserve away from providing individuals and collective capability to support operations, and refocused on 'the raise, train and sustain' functions of ARes headquarters, formations, and units.

Subsequently it was proposed that the HRR category, augmented by the Standby Reserve, would be expanded to support operations, but not the Active Reserve. The Active Reserve, while it would 'remain the core of the Reserve', would have the role to 'raise, train and sustain' the HRR.[561] The HRR was expected to be raised in mid-2006 and key structural changes of units were planned for 2007. One consequence was reported in May 2006 by Major General Barry, that the Reserve Remuneration Review had 'very disappointing outcomes' for the Active Reserves as most of the Budget details were geared to the High Readiness Reserves.[562]

A new Minister for Defence, Dr Brendon J. Nelson MHR (Liberal/Bradfield), in January 2006 looked to further expand the paradigm that Australia's defence interests lay beyond the narrow confines of Australia only, *vs.* the 'Dibb effect' previously accepted by Labor. The HRR was certainly part of the restructuring to meet these developments. In mid-March, a DRA 'task force' led by Major General Barry met with Brigadier Aziz ('Greg') Melick RFD, DGRES-A, for a briefing on the HNA concept and to discuss the impact of removing M113 armoured vehicles from ARes units.[563]

Leading up to this briefing Brigadier McGalliard, the DRA Victoria president, noted to Major General Garde that 'although we accept wholly the concept of HNA, we have doubts about the ability of the Reserves to play their part effectively in it in light of the lack of a number of indicators of action in hand'. McGalliard included how the Active Reserve would provide the HRR and also support the continuation of itself. Garde replied: 'HNA is no panacea as to the fundamental reserve issues that Army needs to tackle.'[564] A subsequent DRA discussion points paper raised a number of serious challenges and issues facing the Active Reserve in delivering the complex 'raise, train and sustain' capabilities required of it by the HNA concept.

In June 2006, the *Australian Financial Review* detailed the formal announcement by the Government of the HRR component. The HRR was to have an initial manning at 1100 men at the end of 2008, and 2800 men and as many as 3500 men by 2012—the Labor Opposition immediately demanded to know why the Government was doing this when it had disbanded the previous Government's Ready Reserve in 1996? In addition, the Government said that it would continue to maintain seven Reserve Response Forces for Australian domestic security, and of course the Active Reserve.

DRA president Major General Barry was noted in the article as concerned that the new HRR would draw numbers out of the Active Reserve, although Major General Wilson, Head of Reserve Policy at Defence, thought that the system would be stable with reservists flexibly

Army Reserve soldiers Sapper Jessica Thomson (left) and Lieutenant Garth Pratten confirm the progress of the security search and lockdown at the Melbourne Cricket Ground prior to the commencement of the 2006 Melbourne Commonwealth Games.
(Defence Images)

moving between the three Reserve components.[565] Separately, ADA executive director Neil James said that the decision to establish the [HRR] was a 'tacit admission that the abolition of the Ready Reserve 10 years ago was a serious mistake.'[566] Some DRA members would agree but many at the time were also unhappy with the Ready Reserve concept because it took in many ex-Regulars.

On 19 August 2006 Major General Barry wrote to CA, Lieutenant General Leahy to consider additional Bushmaster vehicles for the Reserve armoured units to replace the retired M113 vehicles from service. Barry argued that an additional 100 of the Australian-made vehicles would provide ongoing, interesting technical and tactical training as well as operational and especially infantry capability, while giving an extra six company lift to the ADF. These could include the Force Protection Companies being raised as part of the HRR. These training, qualification, maintenance, and movement factors would all provide an additional incentive to retention, which remained an ongoing and serious issue.[567] He also wrote directly to the Minister for Defence, Brendan Nelson, arguing the same points.

In October, Barry received a reply from the Minister, who confirmed that only one of the five Reserve armoured units would receive Bushmasters; the others Land Rovers, as they transitioned to their new light cavalry role under the HNA concept.[568] It was a disappointing response, ignoring the strong capability arguments made by Barry. By the time of the 2006 DRA conference at Randwick Barracks in Sydney on 26 August, the DRA list of issues was clear to invited speakers under the theme of 'Greater Capability for the ADF—A Job for the Reserves'. The list was a useful reminder to all DRA members that there were still 'battles to be fought':

- The HRR is currently being raised.
- The strength of the Army General Reserve is unacceptably low and reaching levels which undermine its ability to support and supply the HRR, let alone provide meaningful augmentation to the Regular forces in other ways.
- There is potential for a similar position to arise in the Air Force.
- The Navy continues to make only minimal use of Reserves.
- The Army qualification and promotion requirements are making it virtually impossible for employed reservists to qualify, be promoted and so feel encouraged to continue to serve. This sees reservists being unable to reach Corporal/Sergeant or officer rank while still spending some time in their units undertaking operational training.
- Command and staff appointments within the Reserve should be for three, not two. This would add to operational capacity, give some stability while not being a threat to the One Army concept.

- For meaningful support to operations Reserve units must be equipped with equipment compatible to that of the regular force.
- There appears to be little realization within Defence that both Regular and Reserve components have different strengths, capacities, and characteristics and that the current 'one solution fits all' is inappropriate and damaging to both components.
- There is still a reluctance to utilise Defence reservists fully, caused by several cultural, organisational, and planning factors.
- The present pay and conditions of reservists remain well below normal community expectations.[569]

One of the ways in which the DRA had changed its approach to the now annual conferences was an even more focused and thorough appreciation of exactly why it was holding a conference in the first place and what objectives it wanted to achieve in doing so. Considerable planning was conducted to ensure that the DRA's message on issues was reflected in the themes and details of presentations. Speakers were carefully chosen to this end and a key audience was no longer the DRA members themselves, but the politicians from both sides who attended as well as the senior ADF officers.[570] For the 2006 conference for example, a theme and program plan was carefully prepared months in advance.[571] The rationale for the 'Greater Capability for the ADF—A Job for the Reserves' theme was summarized as:

- The key issue must be the identification of the required ADF capabilities and the ways the Reserves can contribute to providing them because,
- Once this is accepted by the Government and the Australian Defence Organisation the present state of a continuously uncertain future for the Reserve will be overcome and,
- The maximum use of Reserve resources can be achieved.[572]

This worked both ways of course, as those key audiences for the DRA message quickly recognised. For example, at the end of the conference the VCDF commended the DRA 'for facilitating an excellent forum for announcing measures being taken to maximise the contribution by reservists to the national defence effort.'[573] The conference itself saw papers delivered by a range of senior officers and politicians, including the opening, and closing presentations by the DRA president.

Papers were given by Major General Peter J. Abigail AO, the Director of the Australian Strategic Policy Institute (ASPI); Major General Wilson, ACDF/HRP and Lieutenant General Gillespie VCDF.[574] Politicians who spoke included Senator John A.L. ('Sandy') MacDonald (National/NSW) Parliamentary Secretary to the Minister for Defence, and Robert B. McClelland MHR, Shadow Minister for Defence. Dr Alan Ryan returned following the 2003 conference—he was now referred to as 'the always-controversial' Dr Ryan—and again challenged the Reserves to keep up with complex war-fighting scenarios testing the Regular forces with specialist and combat-ready skill sets.[575]

While there were, apparently, 'poor numbers' attending the conference (60), it was nonetheless declared an outstanding success and a 'conference for the time'. In summarising the conference proceedings, Lieutenant Colonel George, the DRA national secretary, said:

> The standing of the DRA has grown exponentially during the last 12 months due in no small measure to the growing importance of the contribution by reservists to the national defence effort—a contribution without which the ADF could not do the tasks required of it. The DRA is the only organisation outside the chain of command through which the defence message it pertains to reservists can be distributed by Defence and by Government.[576]

The 2006 AGM was held the next day. It was non-controversial, working through routine matters. One branch report, from South Australia, showed that it dropped to only 14 members and 'was struggling' to meet its financial commitments.'[577] Victoria branch noted that it continued to work on policy documents as agreed at the 2005 AGM, expanding some existing ones especially in respect to Navy and Air Force Reserves and Conditions of Service, as well as producing documents on National Service and welfare issues.[578]

Flags at the Sydney Reserve Forces Day 2 July 2006. The 2006 march highlighted those reservists and civilians who have served overseas in the many peacekeeping, disaster relief and other areas of deployment. (Defence Images)

In comments around the RFD March, the branch said: 'This year has seen increased difficulty with some apparently official obstruction and lack of cooperation during preparations for the march'. This needed to be reduced, it said, 'before it becomes too much for the volunteer organisers to manage or tolerate.'[579] The exact same comments were repeated in the annual report a year later—that may have been poor proof reading of the report for that year. One 'quantum leap forward' agreed by all—the DRA web page had gone 'live.'[580]

According to the minutes of the AGM, the AGM was closed by Major General Barry 'in peace, love and harmony', which was a most unusual postscript for any Association meeting of this kind.[581] A paper by Barry subsequent to the AGM was perhaps not so peaceful, as will be seen. The DRA media release following the conference got some coverage with DRA president Barry being extensively quoted in the *Courier Mail*. In the release, 'A Seachange for ADF Reservists', Barry was quoted as saying: 'The Reserves have needed to be used in every Defence operational deployment since 1998. The Government and the Defence Force now understood the need for and the latent capabilities that the Reservist can provide'.

He was critical of the bureaucratic roadblocks that needed to be overcome to obtain the recruits the Government was now seeking:

> The outsourcing of recruiting to a civilian agency, for reservists, had now worked because of the months of delay in processing them and the isolation factor leading to a loss of interest. This was a poor advertisement for Defence, and we must return to our Reserve roots and allow the Services to recruit their own.[582]

The DRA welcomed the introduction of a High Readiness Reserve to enhance Defence readiness with attendant incentives, but 'much more needed to be done with improving terms and conditions'. If Reserves wanted to compete in the competitive labour market, then the lack of superannuation for reservists was a case in point. It was available to every other casual or part-time employee—and the Government was criticised by the DRA for not adhering to its own Work Choices policy.

Another issue, said Barry, was the lack of coordination between Government departments. For example, a service allowance of $10 a day for reservists was announced in the Budget and at the same time the Australian Tax Office decided to take away the *pro rata* Medicare exemption status which reservists had available to them since 1973.[583] In a media release to mark the Federal budget for 2006-2007, Barry remarked that 'Conditions of service of Active Reserves (compared to the new High Readiness Reserves), are so far behind the minimum wage level that no civilian employee could be legally asked to accept such poor conditions.'[584]

Since September 2006, in an incentive to rectify sliding retention rates, Reservist salaries were streamlined with those of the Regular forces as a reflection of overall higher standard of training. One narrative had it that: 'This initiative reflected the change in recent decades,

Captain Connie Jongeneel (L) and Flight Lieutenant Audrey Tan (R) were part of the medical team that in 2006 helped deliver the first baby born at Camp Bradman near Dhanni on the Pakistan side of the Kashmir Line Of Control.
(Defence Images)

that there were now many positions for which little training gap is apparent between reservists and Permanent Force members'.[585] There would have been many in the Reserve who would have disagreed with that statement. As such, the role of the Army Reserve encompassed the '3 Rs'—that is 'reinforcement, round-out and rotation'. With a total strength in 2005-06 of just 15,579 active personnel, recruitment and retention remained an ongoing issue for Defence planners. Nevertheless, reservists continued to have a high training obligation which many struggled to meet, especially if employed.[586]

In an address to the Victorian branch AGM in October 2006, Major General Barry expounded the theme of the need for change at a time when the Regular components of the ADF were recognising the need for the Reserves. He noted that the DRA needed to foster a base for informed comment on Defence Reserve matters and 'had to be pro-active, while maintaining its respect, not being a carping critic'. There was recognition of underlying weaknesses in the Association—it needed a strategic approach, to improve its profile, and work harder with like-minded organisations in all States. He reminded members that the DRA was not a trade union or a welfare organisation but an employee support organisation.[587]

That same month Major General Barry wrote a paper for the DRA committees and members titled: 'Report on the Future of the DRA'. Commenting on the state of the DRA he said that while the DRA had over 1000 members in six States and the Northern Territory, 'we need more to have a positive effect...In earlier iterations some members were seen as Colonel Blimps, troglodytes or dinosaurs and maybe we still do, however I see interested, intelligent, responsible members that want to add value to use of Reserves and the conditions of service for reservists.'[588]

Barry's paper called for an expanded conference to a full two-day format and to further develop the new website which had been launched in August 2006, while noting the improvement in quality in the DRA publication *The Australian Reservist*. But he also charged State branches with 'marking time' and having a 'complacency mindset', offering some suggestions of how this could change—as well as exhorting the members and branches to be more pro-active, and not afraid to put forward new ideas. A call for more numbers was not new—but Barry recognised that numbers meant perceived size and strength.[589]

In footnotes to the end of 2006, Major General Barry visited Canberra in late October for appointments with Commander 2 Division Major General Flawith; national RSL president Major General Bill Crews OA, and with Rear Admiral Graeme S. Shirtley AM RFD RANR and Air Vice Marshal Tony K. Austin AM, both of Defence Health.[590] Following the RFD March launch in Sydney on 2 December 2006 which he attended, Barry commented, 'We obviously have a way to go to have the ADF and politicians understand the importance of a Reservist Parade as a separate event from the permanent forces.'[591] 2007 was to be a more sombre year when the DRA lost three of its stalwarts—Major General Cullen, Association founder, Brigadier Lowen and Colonel Sandow, former Association presidents.

Issue 9 of *The Australian Reservist* was published in February 2007. This was the last edition of journal produced by Lieutenant Colonel Sealey (his wife Beryl was proof reader); he was replaced by an editorial committee, 'not just because of workload but also to inject a tri-service flavour.'[592] Sealey had been given a heart-felt commendation at the 2006 AGM for his hard work over several years to produce an increasingly higher quality journal. The editorial committee consisted of Commander Hicks (technical and Navy), Lieutenant Colonel George (Army), Squadron Leader Michael ('Mike') Dance (Air Force) along with former Lieutenant Colonel John Morkham and Major Bedggood.[593]

Development of a DRA website (www.dra.org.au) was well underway by this time as the DRA now recognised the need to keep up with the rapid changes in public communications around the internet and digital media—the first rendition of the new DRA website was launched on 15 August 2006. DRA Western Australia branch had proudly reported to the August 2006 AGM that it already had had a website for nearly nine years although it had been re-designed by a professional just two months prior.[594] One wonders how much more effective Western Australia branch had been by being an 'early adopter' of the new technology, even without a professional design, compared to the later DRA National Executive's arrival on the scene.

DRA focus in the early months of 2007, while occupied with the running dialogue over the loss of the Medicare levy exemption, the HRR and the Minister for Defence 12-month Gap proposal, was on the most important project of its national submission to the Military Superannuation Review initiated by Minister for veteran Affairs Bruce F. Billson MHR (Liberal/Dunkley). The DRA saw the submission as 'a real opportunity to have the Government and the ADF address this anomalous situation.'[595] The submission, prepared on behalf of the Association by Major General Garde, ended with the statement: 'It is high time for reservists to have the benefit of a superannuation scheme and it is now ten years since superannuation contributions were mandated of civil employers.'[596]

Other work of the DRA continued on through the year. In late May, Major General Barry visited Canberra to meet Parliamentary Secretary to the Minister of Defence Peter Lindsay, the ASPI executive director, Major General Peter Abigail AO, and ADA executive director, Neil James.[597] In late July 2007, the DRA finalised its national submission on the *Defence Reserves Service (Protection) Act*.[598]

The passing of Major General Paul Cullen on 7 October 2007, the same day as the DRA AGM that year, was met with dismay even though at the age of 98 it was not unexpected. His military career of more than 40 years saw him rise through active service during World War Two to CMF Member of the Military Board and various and well-deserved honours both civil and military. In his civil life, among his many business and philanthropic milestones, he founded Mainguard merchant bank and Austcare and was a successful cattle breeder. As Cullen's obituary in *The Australian Reservist*, written by his long standing friend and colleague Major General Glenny, described him, Cullen was 'a humanitarian warrior who devoted his life to service.'[599] In the context of the Reserves, however, it was his role as founder and leader of the CMF Association in 1970 which is, as the DRA, his lasting legacy.

In this role and against considerable opposition, especially at first, Cullen remained a determined advocate of his beloved Reserves, always focused on achieving positive outcomes

Private David Laube, from Renmark, SA with the 10/27 Battalion, RSAR moves upstairs inside Sydney Town Hall to continue low level searches with the rest of his team as part of Operation DELUGE, providing low risk search and security checks for the Asia Pacific Economic Cooperation (APEC) meetings.
(Defence Images)

for Reserves organisational as well as welfare levels. He had no hesitation in direct engagement with Prime Ministers, senior ministers, bureaucrats, senior regular and other service officers in that advocacy. Exasperating to some, vexing to others, Cullen nonetheless remained steadfast to his ideals and did not hesitate to expose cant and flawed thinking which could adversely affect the Reserves. Cullen attended DRA conferences almost to the end of his life and continued to push reservists in the DRA to keep thinking about and taking action to find ways to deliver better outcomes for the Reserves. He was a most remarkable man. He was accorded a military funeral at Victoria Barracks in Sydney.

The 2007 DRA conference was the third to be based on a theme 'to raise the level of debate and action on the role of the Reserve forces within the ADF, with particular emphasis on enhancing their capability'. The conference theme was 'ADF Reserves—Present Opportunities, Future Realities'.[600] The conference attracted a range of influential speakers, including ACDF-R/HRP, Major General Greg Melick AM RFD and the

Deputy Commander of Joint Force Headquarters, Rear Admiral Rowan C. Moffitt AM RAN, while Major General Garde spoke to the DRA submission to the Military Superannuation Review and on the review of the Protection Legislation.

Speakers included the Directors-General of the three service Reserve components; the Surgeon-General Rear Admiral Shirtley; the Director of ADF Recruiting, Captain Cameron G. McCracken RAN; the Director of Capability, Operations and Plans, Colonel Neil Greet, and also Dr H. Smith from UNSW on the review of the Defence Reserves which had not been made public at the time of the conference.[601] Senator Marise A. Payne (Liberal/NSW) a future Minister for Defence, gave the Government view on Defence, although she offered no specific initiatives to support Reserves.

Major General Barry summarised the dire situation for Reserves prior to the East Timor events of 1999, and the rapid changes thereafter as the operations in East Timor 'became the 'watershed' for the ADF, highlighting all the deficiencies in its structure and function when it came to sustained operations…it created a new political awareness which led to the 2000 White Paper…'. The Reserve Forces in turn moved from their pre-1999 structural expansion base through a defence of homeland vital assets, to an intermediate operational capability in 2007, 'now gathering momentum because Reserve Forces are now needed more than ever to augment the ADF'. Barry finished his overview by noting that Conditions of Service had not kept pace with the organisational and operational demands now placed on the Reserves, and this was the priority for the DRA in the months ahead.[602]

Regarding relationships with other organisations, Major General Barry recorded that Major General Glenny was the DRA representative on the national council of the DRSC ('we need to work closely with this Government sponsored Council'); Colonel Haynes was the DRA representative on the national council of the Australian Veterans and Defence Council Services Council (AVADSC) and that there were DRA representatives on the NSW, Tasmanian and Victorian councils of that body as well.

With the demise of the Armed Forces Federation of Australia (AFFA), the Defence Force Welfare Association (DFWA) has assumed its welfare role and the DRA had made contact with and needed to cooperate with DFWA. Barry also commented that the 'recent development in profile…and direct liaison with the ADA executive director, Neil James, 'has proven to be most worthwhile as he continually refers Reserve matters to us'. Barry also reported that although there was no direct linkage with the RUSI 'we have a large cross over of membership and this suits both organisations'.[603]

In terms of the Association itself, Major General Barry reported on developments in his 2007 annual report for the AGM that year. In his accompanying executive circular to the branch committees at the same time he said that the initial review of DRA policies was now complete, for discussion of the next phase at the AGM.[604] He noted in his annual report that 2005 and 2006 amendments to the Association's constitution had also been approved by State branches. Barry regarded conditions of service for reservists as the Association's priority with emphasis on the right to superannuation and more aligned remuneration with the Regular component. He called for a wider discussion of Tasmania's Tasman Scheme and considered that it 'needs to go national'.

Barry also noted that the Association needed to address financial stability and operational expenditure at the forthcoming AGM 'if we are to progress effectively with our advocacy'. Related to his point was a call for increased membership to provide for increased financial flow.[605] This was by no means the first call by an Association president or branch presidents for increased membership over the decades of the Association's existence. Major General Barry made the point however that while sponsorship should continue to be an objective, 'we need to tap into our membership to be self-sufficient'.[606] Membership numbers was as much about sustainability as it was about profile.

The Branch reports for the 2007 AGM give an interesting overview of the health and vitality of the Association that year. Queensland reported 108 members (of which only 50 were financial, along with six life members), with 92% from Army, 32% serving in the Active Reserve, 13% in the Standby Reserve and the remaining 49% retired members. This was a far cry from the beginnings of the CMF Association when serving members viewed the new association with considerable suspicion and held it at arm's length. Of the remaining eight percent of membership overall, four percent were Navy, of which 50% were serving, and four percent were Air Force, with 75% of those still serving. The branch noted that its committee was meeting every six weeks, and held three seminars during the year, along with an annual dinner, supported the RFD activities, produced two newsletters and other communications over the past year. The one submission made to the National Executive of the DRA was on the superannuation question.[607]

The WA branch, by contrast, reported 216 members (including 88 life members) and of these 28 were serving. However, its activities it seems were constrained to matters affecting local recruiting and retention and support for RFD. With only 13 Brigade as the major Army Reserve formation in the State and virtually no RAN and RAAF Reserve presence—no tri-service membership was even reported, indicating that it was not significant—it was perhaps unsurprising that national activities at least that year were muted despite the large membership. It is not known how many of the membership were financial in that year.[608]

As could be anticipated, the NSW report was among the most comprehensive. The branch had focused on military compensation over the past year along with other issues in line with the focus of the National Executive of the DRA. The branch had experienced a drop in membership. However, that was due in main part to a deliberate effort to cull non-financial members, yet the branch still managed to have 282 members. It acknowledged that serving members of the Reserve were the key to future membership and efforts to do attract them would be continued but admitted that it 'struggled' to gain them as new members.

NSW branch was still putting in much time towards supporting RFD and working with kindred organisations, while focused on the Military Compensation Scheme and supporting commanding officers to write submissions for Reserve honours and awards. The Branch also reported on the success of the 'schools programme run over the previous three years' to increase the awareness of Australia's military history in the NSW school system. This appears in hindsight to be a type of 'mission creep' which almost certainly detracted from a focus on the main branch, and DRA, objectives.[609]

One curious comment in the NSW report was that meetings were conducted with commanders 'in an endeavour to alter the current innocuous thought process that seems to want to support failure. We can only keep trying to effect that policy that we all know is not in the interest of anything bar immediacy.'[610] Was this an indication of the frustration, seen before many times over the Association's existence, between retired senior officers and serving officers in the chain of command, albeit both in the Reserve 'family'?

The branch, according to the report, was now putting efforts 'into trying to reverse some of the pathetic decisions that have been made in the recent past that we believe is destroying our Reserve Force'. Colonel Fleeton, the NSW president was adamant: 'Unless we can be part of the influencing group that can reverse the downward spiral, the Reserve is finished, and the ongoing defence of our nation would be at risk.'[611] While the language was strong, there was no doubting the sentiment. Whether the DRA, let alone branches, could actually reverse decisions was another matter. It was a rare achievement when this happened.

Last, the Victorian branch report for the 2007 AGM reflected its role as a *de facto* secretariat for the Melbourne based DRA National Executive, with its subsequent involvement in furthering policy writing and development for the DRA.[612] With 230 members and a larger committee than other branches, with monthly meetings and *ad hoc* sub-committee meetings,

Lieutenant General Peter Leahy AO Chief of Army joined ADF personnel from many of Victoria's Reserve units, including past National Servicemen, to celebrate Reserve Forces Day in Melbourne on 2 July 2007.
(Defence Images)

it was active on several fronts. Like other branches it continued to be deeply involved in the annual RFD (now named the Reserve Forces Support Day—RFSD—at least in Victoria), and in its cooperation with kindred organisations.

In comments around the RFSD the branch noted that the 'virtual absence of serving Army reservists' (compared to the RAN and RAAF presence) 'was particularly noticeable'.[613] In January 2008 Major General Garde succeeded Brigadier McGalliard as branch president with a Navy and Air Force vice-president in support. McGalliard had served for ten years as branch president as well as seven years as national vice-president; he was awarded OA for his service to Defence Reserves and the DRA.

The election of the Labor government under PM Kevin M. Rudd MHR (Labor/Griffith), in November 2007 brought in new thinking and more changes for the Reserve. Joel Fitzgibbon MP (Labor/Hunter) became Minister for Defence.[614] Major General Barry, meanwhile, had taken to his role as president with gusto and was energetic in pursuing opportunities to show the DRA flag. Following the change of Government, the DRA was quick to acknowledge the new Defence appointments of Labor and follow up with visits to Canberra. The DRA had been relatively assiduous in maintaining contact with Labor 'shadows' while they were in Opposition, along with the ACTU, and inviting opposition 'shadows' to attend the DRA conferences.

This allowed the DRA to maintain its momentum on Reserve issues. In May 2008 for example, Barry spent quality time in Canberra, meeting the Parliamentary Secretary to the Minister for Defence, Dr Michael ('Mike') J. Kelly MHR (Labor/Eden-Monaro—a former Army Legal Service colonel), Opposition spokesmen Peter Lindsay MP and Senator Nicholas ('Nick') H. Minchin (Liberal/SA), and the ACTU representative on the DRSC, among a range of other Defence and kindred ex-service organisations.[615] New blood entered the DRA scene with the election of Major General Irving as national vice-president in place of Brigadier McGalliard who had served in that role for 10 years. Captain Peter W. Wertheimer RFD replaced Lieutenant Colonel Sharkey who had stepped down as national secretary during the year.[616]

An update of RAAF Reserve came in an article in *The Australian Reservist* in early 2008, 'Integrating the AF Reserve with the Regular Force', by the Deputy Director General Reserves—Air Force, Group Captain Carl F. Schiller OAM. Schiller detailed the recent RAAF Reserve Restructure which saw the integration of the RAAF Reserve into Permanent Air Force units to focus Reserve service on meeting specific wartime tasks. The Reserve Restructure was described as 'a landmark event' and 'the most significant change to the Air Force Reserves since the 1960s'.[617] The restructure included a new High Readiness Reserve and placed responsibility for Reserve training onto the Permanent Air Force. From the DRA's perspective, while the RAAF Reserve remained a small component overall in the ADF, the DRA's emphasis on equitable Conditions of Service was no less important.

The 2008 DRA Conference and AGM was held over 22-24 August that year. The conference was the first at a non-Army base, i.e. at historic Point Cook, the former RAAF base, now known with Laverton, as RAAF Williams. Continuing the theme-based approach for conferences, the 2008 conference looked at the ways to enhance ADF capability by appropriate use of reservists as cost effective assets. The conference was held in the knowledge that the Government had commissioned a new Defence White Paper with public comment to be made by 1 October. The Government speaker defined three distinct Reserve models, namely the citizen Reservist (Army model), former permanent force members (Air Force/Navy) and the specialist (health, legal etc.) across all three services. The conference was asked to discuss how each group could contribute meaningful capability within the White Paper framework of tightening budget and unpredictable national security risks; the Government had already started to cut Reserve training days.

Other presentations included 'ADF Reserves—Up to and beyond the White Paper', by Major General Melick, Head Reserve and Employer Support Division; 'Increasing Reserve Capability' by Captain Richard Phillips, RANR, Director Navy Reserve Support—

National; 'Army Reserve 2008—Achievements and the Future' by Brigadier Bruce Cook—Director-General Reserves—Army; 'Issues for the RAAF Reserve' by Air Commodore Peter J. McDermott AM CSC, Director-General Reserves—Air Force; and 'The Employer Support Payment Scheme' by Squadron Leader Gilbertson, HQ Reserve Training Wing. Rear Admiral Shirtley also apprised the conference of a new development and long overdue major reform—the transformation of Defence Health Services into a Joint Health Command.

The president, Major General Barry reported to the subsequent AGM that a number of reviews were on hold pending the 2008 Defence White Paper, including the Review of Reserve Capability by HRES Division, the Reserve Remuneration Review, and the Superannuation submission with kindred organisation DFWA, and the review of the *Defence Reserve Service (Protection) Act 2001*. It was anticipated that DRA policies, being reviewed by Brigadier McGalliard and others in Victoria for further branch discussion, would be looked at again once the 2008 White paper was released.[618]

On 16 September 2008 Barry was able to present the DRA position paper to the Defence White Paper Consultation Panel. By February 2009, the White Paper had still not been delivered and was expected in April of that year. The paper was being 'conflicted between maintaining and developing capability *vs*. effecting savings in the economic climate (thus the character of the White Paper is changing the further it is delayed because of the impact of economics and changing threats due to the changing economic situation)'.[619]

In his 2008 annual report, Barry again called for ideas to gain sponsorship, to increase membership and to develop 'succession plans to obtain fresh ideas'. Regarding sponsorship, Barry's logic was that first the Association had to be able to support itself, but without funding from all sources, how could the DRA maintain a presence in Canberra and be an effective advocacy organisation? Membership of the Association had remained stable in branches for some time, especially Tasmania and Victoria, but South Australia continued to struggle and numbers in NSW had dropped by over 100 since 2001.[620] Meanwhile in late 2008 a company from 1 Commando Regiment became the first formed Army Reserve unit to see combat since World War II when it was deployed to Afghanistan. 'The initial deployment proved problematic however, with a subsequent inquiry finding that the company had received less support for its pre-deployment preparations than was typical for regular units and that its training was inadequate'.[621]

One of the contributions to thinking on the Reserves at the time was a widely circulated paper written by Drs H. Smith and Brigadier Jans—'Use them or lose them? Australia's

Defence Force Reserves', first published in November 2008 by ASPI. The paper concluded that 'there was a viable and meaningful role for the reserves...provided that the ADF and government are prepared to devote sufficient effort to them and assign them appropriate tasks'. [622] The DRA was quick to pick up on the 'Use them or lose them' catch-cry. This was as busy a time for the DRA as ever before, but 2009 began on a sombre note, with the news that Private Gregory M. Sher, a Reservist on full time duty with 1 Commando Regiment had been killed in action in Afghanistan. He was the first Reservist killed in action since World War II.

The March 2009 issue of *The Australian Reservist* saw DRA president Major General Barry extoll the delivery by the Reserves of considerable support to Victorian authorities fighting large scale bushfires in that State. At the same time, Barry noted to DRA members that: 'The Association has been somewhat frustrated over the past 12 months in its advocacy for equity in terms and conditions of service for reservists'. The DRA now had to await the outcomes of the Defence White Paper now anticipated in May 2009 with the Government's budget. The budget had already been impacted by what became known as the GFC—the Global Financial Crisis—which had deeply, if temporarily, heavily impacted on the Australian economy. The DRA could only wait and see.

Barry had replaced Major General Glenny as the DRA representative on the DRSC which was 'growing in importance in its role of communicating the Defence reserves message to employers'. He had also accepted the position of Vice President—Reserves on the DFWA which gave the DRA 'the opportunity to better participate in and collaborate with the work of other important Ex-Service Organisations (ESOs)'.[623] The DRA journal showed the work of its editorial committee with a diverse series of articles. These included policy updates from new DRA vice president Major General Irving; a paper reflecting on 'Improving Capability through the Reserve' by Dr Andrew Davies—Director of Operations and Capability, ASPI; a profile of the new Surgeon-General Australian Defence Health Reserves, Major General Jeffrey V. Rosenfeld, OBE, RAAMC MD; and 'Making the Most of the Naval Reserve' by Captain Joseph L. Lukaitis RFD. The depth of the journal was another sign of the increasing sophistication of the DRA's main publication.

Reserve Forces Day, at least in Victoria, came under scrutiny. The Victorian president, Major General Garde, led the discussion on the future of RFD with an article in the branch newsletter in March 2009. Garde commented that since the inception of RFD in 1998, the DRA had always been at the forefront of volunteer efforts to keep RFD among the more important of its efforts to support the Reserves. However, he now believed that 'RFD in its present form is not seen as usefully advancing the cause of reserves'.

There were several reasons for this. Support for RFD had dried up among Reserve COs, essentially because of financial cutbacks across the board. Increasing numbers of reservists had been deployed operationally; they preferred to support Anzac Day, RFD was not seen as relevant to them. RFD could not attract the financial resources to continue, and the combination of these reasons and other identified obstacles meant that RFD in 2009 was under threat as perhaps the last event of its kind, especially with preparations for the centenary of WWI approaching. In short, Garde said, 'RFD must modernise and reposition if it is to succeed.'[624]

Meanwhile the Government's White Paper, 'Defending Australia in the Asia Pacific Century: Force 2030', was released, '...based around assumptions that China would become increasingly dominant in Australia's region and that Australia cannot rely on the United States for protection.'[625] One change to the language used for Reserves in the White Paper was noticeable. Now it was 'part-time' and full-time members of the ADF. An article in *The Weekend Australian* in May 2009 by Mark Dodd, 'Army Reserve weak link in defence plans', highlighted that continued falls in Reserve numbers 'cast doubts over ambitious defence white paper plans to better integrate the part-timers into the army's full-time ranks'. Reserve numbers for Army were only 15,400 or 71 per cent of its authorized strength. The article was based on the Auditor-General Report into the Army Reserve, released on 8 October. DRA Victoria called the report a 'reality check and an authoritative statement.'[626]

The DRA Conference in 2009, once again held at RAAF Williams (Laverton) in Victoria in August that year, was devoted entirely to 'The 2009 Defence White Paper—The Reserves toward 2014', with a focus on the first five years rather than the full range of the White Paper to 2030. The conference included political input from both Government and Opposition, evaluations of the White Paper itself and inputs on the implementation by the three Services. Papers included: 'Implications of the White Paper for the Army Reserve' by Major General John G. Caligari DSC AM; 'Enhancing Navy Capability' by Commodore Ranford Elsey RFD RANR, Director-General Reserves—Navy; 'Update on the Status of the Army Reserve' by Brigadier B. Cook, Director-General Reserves—Army and 'The Defence White Paper 2009—the Opposition View' by Peter Lindsay, Shadow Parliamentary Secretary for Defence.

The various presentations gave mixed reports on the White Paper re the Reserves. While noting the long-term horizon and the aspirational equipment targets (and paucity of related costings), Major General Barry believed that it also showed 'the lack of any normalization of conditions of service for reservists not on continuous full-time service. This...omission', said Barry, 'highlights the DRA's prime objective of reservists obtaining more compatible conditions of service to their permanent counterparts'.[627] Although Navy and Air Force had delivered positive papers at the Conference around ongoing integration, the restructure of the Army Reserve was still a work in progress and a review was to be undertaken. The DRA president said: 'The Army Reserve should not fear change if it is appropriate', but in the meantime set up meetings with CDF, VCDF and CA 'to express concerns of many as to the process of the proposed restructure and the lack of progress regarding appropriate conditions of service for all reservists'.[628]

At the AGM following the conference Major General Barry reported on the activities that he had been involved in over the past year. Perhaps unsurprisingly, given that review schedules rarely, if ever, kept to their original timetables, Barry reported that the Review of Reserve Capability by HRES Division may have been shelved (it was), along with the Reserve Remuneration Review. The *Defence Reserve Service (Protection) Act 2001* Review remained 'pending'. Barry noted that DRA policies, updated in 2007/2008, now had to be reviewed again light of the 2009 Defence White Paper, a reminder that 'DRA policies needed to be dynamic' and that "We need to grow and reinvigorate ourselves."[629] By this time in Victoria, Reserve Forces Support Day had further morphed into the Defence Reserves Support Day (DRSD); in other States it remained as Reserve Forces Day.

The Victorian branch noted that the Tasman Scheme, which the DRA supported nationally, and which had got underway in Victoria, had still not been implemented eight months after selection of the junior NCOs for deployment to training in New Zealand.[630] In changes to the DRA constitution, came succession planning with two year, not one-year terms for key officers and four new vice-president positions. This led to the appointments of Major General Flawith as VP-Army, Air Commodore Peter McDermott AM CSC as VP-Air Force and Rear Admiral Shirtley as VP-Defence Health. The VP-Navy position was left vacant but was filled in 2010 by Commodore Elsey.

The NSW report to the AGM included the branch president's remarks that forthcoming changes to support Reserve service more fully 'may even take us away from the 'Woolworths' approach ...the shopping for individual [Reserve] members to round out vacancies within ARA positions leaves the Reserve unit short of special people and does nothing for the long term effectiveness of the unit'.[631] South Australia reported a good year with 47 individual members on its books, an all-time high, with Major General Wilson succeeding long-standing president Lieutenant Colonel Alexander. SA branch had also selected candidates for the Tasman Scheme but reflected Victoria's concern about the lengthy delay to actual deployment.[632]

Tasmania branch maintained a high membership of 161 'life' and 24 individual members that year. The branch's Tasman scheme selections had similarly stalled—in Reserve Support Branch—although no reasons were indicated in the report. Overall Tasmania was focused on working with kindred and ex-service organizations, but it did comment on recruiting and ability to train recruits as major issues in its region. It also noted that the build-up from ex-permanent members—who only have to commit to the Standby Reserve—had to date not been successful; 'all this has done is inflate numbers of non-committed reservists.'[633] The DRA Western Australia branch also reported high member numbers of 173 including 93 'life' members but noted there had been no RFD March due to Chief of Army policy.[634]

In 2008–09 total strength included 17,064 active personnel. In addition, there were another 12,496 members of the Standby Reserve.[635] Throughout 2009 and into early 2010 there had been a consistent media to-and-fro between critics of the pressures being exerted on the Reserves, the supporters of the Reserves such as the DRA, and a Government and Defence on the back foot. News of yet another review of the Reserve which was supposed to go to Government before the end of 2009 started the sequence of events, with reports that budget cuts would affect the Reserves significantly and that many employers were dissatisfied with multiple deployments of their Reserve employees as the tempo of F/T Reserve service increased. Funding cuts affected ammunition stocks and training days were further reduced for the Active Reserve, along with participation in such events as Anzac Day.[636]

In December 2009 DRA president Major General Barry issued a press release 'Reserves Hit for Six by Army Leadership', which decried the impact of the Army Reserve Restructure Plan and detailing six areas in which the proposed plans would adversely affect Reserve capability. This was quickly followed by a radio interview with the ABC to discuss these and other issues such as cutbacks in Reserve salaries, adverse morale consequences and likely future costs of increased short-term separations. A press article in the *West Australia News* also made similar references to the impacts on Reserves in WA. Another article in the *Canberra Times* (which cited Barry) also came out before the end of the year. The Government and the CA, Lieutenant General Ken Gillespie, responded with press articles and letters to the editor to 'set the record straight.'[637]

Calls for 'realistic and professional management of the Defence Reserves' came in a letter to the *Canberra Times* from DRA secretary Lieutenant Colonel Ian George along with a letter to the editor from the ADA decrying training day cutbacks, both published in early January. Other articles in the *Herald Sun* in Victoria and on the ABC continued the flurry of public criticism of the Government and Army. Major General Garde commented:

It is as if there has been a sudden awakening that Reserves provide a valuable operational asset. Given the operational importance of Reserves one might be surprised to find that the Reserves are still operating under a 20% training cutback...Capability will not be advanced unless it is properly resourced and equipped. As a key objective is to withdraw half of the existing regular manning from reserve units, I am expecting that the organizational changes will be savage.[638]

In the next issue of *The Australian Reservist* in March 2010, Major General Barry noted that since the 2009 conference, 'Army has been left in a quandary of uncertainty fed by substantive rumours....nothing structurally dramatic has changed for the ARes except the savage cuts to ARes training days to ensure salary payments to their regular counterparts, who are overstrength by some 1100 personnel'. Barry reported that the meetings held on 3 September 2009 between himself, Major General Garde and Major General Wilson with the CDF, VCDF and CA had done nothing to allay their concerns other than, subsequently, being told there would be no restructure of the ARes before the next Federal election [in 2010].[639] If there was to be a restructure certainly the ARes wanted to be consulted.

The July newsletter of DRA Victoria noted changes to Navy and Air Force Reserves. Navy had recently cut its Active Reserves by about a third of its strength with a reduction in training days for those that remain. Air Force has integrated Reserve Squadrons at the same

A Special Forces Reservist watches his colleagues enter a Blackhawk in Uruzgan. SOTG XIV, based on a company group drawn from Army reservists of 1 Commando Regiment, partnered with the Provincial Response Company Uruzgan (PRC-U) in operations to support the Afghan government, 2010.
(Defence Images)

time reducing the total Reserve manning. With a Federal election pending, Major General Garde hoped that 'we are given Defence Ministers with wisdom and judgement. The incoming Government will have to make serious defence decisions including as to the capabilities and future of Reserves in all three services'.[640] Meanwhile cutbacks to training continued along with criticism of Defence and the Government for reported plans to cut regular cadre in Reserve units and disband other units altogether, including university regiments. Those plans had been reported by Ken Aldred, a former Liberal MP and Reserve officer and a long-time supporter of the Reserves.[641]

In his annual report for 2009/10, DRA NSW president Colonel Fleeton recorded his view that Government should separate the Defence budget into the full time and Reserve components which 'may then put to bed the nonsense that occurs each year [when] training days are cut'. Fleeton related a discussion with a young soldier who was the only one of 23 who had been on his recruit course, saying that most had left due to having no days to train. 'Surely', Fleeton said, 'the warning bells would be ringing and that this waste of funds will cease.'[642] It seems not; but Fleeton's complaint was a common concern throughout the DRA. The NSW branch had been given full membership at the State's DVA Deputy Commissioner's Consultative Forum, a useful forum for Colonel Fleeton to 'educate' the deputy commissioner and her staff on Reserve matters. With 149 members, the NSW branch was busy with a wide range of activities over the past year under review.[643]

A Federal election was called for the same day as the DRA had planned for its conference—on 21 August 2010. The conference was postponed and was finally held at Laverton in Victoria once again, over 23/24 October. The theme was 'ADF Reserves, Strategic Direction for the Future'. In preparing for the conference Brigadier McGalliard circulated discussion papers and reports to focus DRA planners on the conference detail.

In April McGalliard noted that the current Defence/ARes/DRA situation—an uncertain future in the shadow of the forthcoming Federal election—'arises from a mix of defence theory, strategy, political attitudes, departmental financial brinksmanship, short term thinking, long term prejudice and a certain amount of bloody-mindedness'.[644] The conference was designed to reflect on the constraints being placed on timing and planning cycles by one-year budget horizons or at best three-year electoral cycles, whereas the Reserves needed more like a ten-year future. The conclusion was that current operational demands at the time were important but not the sole criteria of planning. So, the conference in 2010 looked at force types and needs which might be required in the future vs. a range of immediate times and needs.[645]

Major General Garde gave the keynote address, outlining the issues which have arisen through, or because of, the 2009 Defence White Paper (DWP'09) along with its Strategic Reform Program (SRP). In a major coup, the conference was able to have Colonel Charles G. Ikins USMCR (Rtd) present a paper for US Assistant Secretary of Reserve Affairs, Dennis M. McCarthy, on how the US structures and uses its Reserves. A range of other quality speakers made presentations. 'ADF Reserves—Their Future—The Government View' was presented by Senator David I. Feeney (Labor/Victoria), the Government's new Parliamentary Secretary for Defence.

The Opposition view was presented by Senator Gary J. J. Humphries (Liberal/ACT), Shadow Parliamentary Secretary for Defence Materiel) representing the Shadow Minister for Defence Science, Technology and Personnel. Brigadier Jans from the Centre for Defence Leadership and Ethics at the Australian Defence College (ADC), spoke to his earlier paper 'Use Them or Lose Them—Australia's Defence Force Reserves' with Dr H. Smith. Report cards were given on the current situation with Reserves by CRESD, Major General Melick and the three service directors-general Commodore Elsey, Brigadier Iain Spence CSC RFD and Air Commodore Terence ('Terry') C. Delahunty AM, along with Major General Rosenfeld, the Surgeon-General Health Reserves. The VCDF, Lieutenant General David J. Hurley AC DSC, was also present.

The end of 2010 approached with much uncertainty. Major General Barry reported to members in October 2010:

> The DWP'09 is seen by many to be altruistic and so far there have been no published outcomes...though the rumoured changes for the Army Reserve have not been well received. ....The SRP [Strategic Reform Program] is about obtaining efficiencies... the outcomes however for Navy and Air Force Reserves have seen them substantially downsized, whilst Army seems to have strong aspirations to downsize their Reserves all at a time when reservists are under increasing demands for their operational service!
>
> The 'cause' for downsizing is a consequence of the amounts set aside...for payment of personnel salaries...with permanent forces overstrength and separation rates at all-time lows the only 'wriggle room' is for Reservist's activity to be reduced; in some cases drastically and in particular a reduction in the number of reserve personnel undertaking continuous f/t service.[646]

Barry concluded: 'It is a challenge for the DRA to advocate on behalf of all Australians that believe in having an effective Defence Force with an efficient and effective Reserve element.'[647] By the end of 2010 Navy Reserve was down to 600 personnel—'it has virtually ceased to exist.'[648] What would happen to the remainder, and could the DRA keep the Government and

Defence to its promises regarding the Reserve? Another decade beckoned. Nearly 30 years earlier, a series of articles about the declining state of the Reserves in *Defender*, the magazine of the Australian Defence Association, included one titled 'Going-Going-G---'.[649]

It wasn't the first time that the future of the Reserves was questioned, or someone had predicted its demise. For the DRA, the prime advocate and supporter of a sustainable Reserve, the next decade would bring further challenges. The irony, for those veterans of the Reserve and the DRA since it began as the CMF Association in 1970, was so many of these challenges had been faced before. For many, 'the more things changed, the more they remained the same'. It seemed, for others, that the ADF was a tiger chewing on its own Reserve tail. The character and traditions of the original CMF had now changed almost beyond recognition. The DRA had been forced to adapt as well.

11.
Plans BEERSHEBA and SUAKIN

With *The Australian Reservist* now being consistently published twice a year compared to some earlier years, the edition in April 2011 tended to be full of interesting content from specific articles around reservists and service with the Defence Reserves to book reviews, travelogues, history, and Reserve activities both home and abroad. That is not to say that it had become Army or Navy or Air Force news 'light', as there were always one or two articles around policy or conditions of service for Reserves. For example, the April 2011 journal, now more than 40 pages long, included articles such as 'Collecting the Civil Skills of the ADF Reserve' and 'Employer Support Vital for Mission Success'. However, it also gave an opportunity for more junior ranking officers and NCOs to join in the publication and in a very positive way, demonstrate the diversity of views within the Reserve and not just reflect the views of senior officers.

The later edition of the journal, usually in September following the annual conference and AGM, while still containing a range of articles as described above, devoted about half of the space to selected national conference presentations. In both editions, the DRA president's message remained an important medium for the president to reach out and update committees and members at large on broader issues affecting the Reserves. The April 2011 president message was an especially important start to the decade, for Major General Jim Barry reviewed developments since November 2010, developments which reflected the recognition of both the importance of the Reserves to the Total Army but also of the DRA itself as an organisation that had to be consulted around change.

First, Barry reported that the aims of the conference the previous October had been more than met. The Army's draft restructuring plan for the Reserves by reducing Regular Army cadre had been curtailed. The conference had got the point across that the Reserve was not simply an organisation of convenience which could be switched on or off or adjusted at every financial whim, but had its proper and indeed, critical place and contribution to make within the context of DWP'09. Major General Greg Garde's keynote address had been quite influential, noting the ten strategic issues that must be considered before major restructuring was considered or implemented. While Army had managed to retain its numbers overall, Air Force and Navy Reserves had been substantially downsized due to financial pressures above all else.[650]

Major General Barry went on to relate how, during his visit to Canberra in November 2010, and at the DRA's request, he had received a briefing from the 'new' Cadet, Reserve and Employer Support Division (CRESD) regarding its progress of work on the so-called 'Reserve Reform Stream', as part of the Strategic Reform Program (SRP).[651] The SRP in turn was running parallel with Army's 'Force Modernisation Reviews' arising from the DWP'09.

Through this briefing, Barry became aware of 'Plan BEERSHEBA'. Although not formally announced until December 2011, this Plan was a significant restructure of the Australian Army, and indeed the Reserves. The process of implementing the organisational changes began in 2014, and it was not completed until 2017.[652] Barry immediately offered the DRA to assist in the consultation and communication with the Reserve community through a 'sounding board' of senior former commanders to act as a review panel—these were Major Generals Barry, Garde, Irving, and Wilson.[653] The offer was accepted by Army.

Corporal Karl Tabuai, 27, an Army Reservist with 51 Battalion, Far North Queensland Regiment, training in the Shoalwater Bay Training Area during *Exercise Talisman Sabre* 2011.
(Defence Images)

Subsequently, in December 2010 the panel was briefed in the Melbourne office of Ernst & Young, which was the consulting firm engaged tom develop the cost/benefit of the Reservist model hosted by CRESD. Two complementary projects were of interest—the Civil Skills database (launched by the CDF in March 2011 and addressed in detail in the April journal) and the study into the Conditions of Service for Reserves against capability requirements. The DRA panel received a second briefing from Ernst & Young in March 2011. As Barry said in his report, 'This consultative approach is unprecedented in my experience' ...for the first time ..."we are inside the tent" and therefore will understand the rationale behind the decisions.'[654]

In February 2011, the DRA responded for a request for a submission to the outcomes of the first of series of Reserve Modernisation Workshops (RMWs) which had been held that month for serving one-star officers, followed by a second round in mid-April 2011. The April RMW examined ARes training, administration and governance, and command and control at Brigade level, but on this occasion the outcomes were not publicised, nor was the DRA asked to make submissions in response. With the DRA national conference scheduled for 19-21 August 2011 at Keswick Barracks, Adelaide, the timing was exceptional as it would allow the restructuring plans of all three services to be presented. As Barry commented, 'The key element to the restructure is that the Reserve must be considered within the Total Force Concept and not as a "bolt-on" or worse still, as an afterthought'.

The pace of change gained momentum when the RMWs were overtaken by restarted Force Modernisation Reviews to develop the Army's structure. Milestones were endorsed, by the CDF by September and finally, by the Minister for Defence in October 2011. In the meantime, the DRA conference went ahead in Adelaide under the theme of 'ADF Reserves Capability—Where to Now?'.

Leading up to the conference was a meeting of an Army Modernisation Steering Group (Enhanced) on 11 July and the outcomes from that meeting were reflected in DRA conference proceedings the following month. Selected papers from the conference were provided in the September edition of *The Australian Reservist*. In the 2010 conference a comparison with the US Reserves had been made; at the 2011 conference the Canadian Reserves structure was examined, through a presentation by Rear Admiral Jennifer Bennett OMM CD, the Canadian Chief of Reserves and Cadets. Her overview was a highlight for conference attendees as it especially allowed the DRA to compare the Canadian Naval Reserve with the RAN Reserve. The conference focused heavily on Army and Navy Reserves with Air Force Reserves to be looked at in 2012.

The Government view of the ADF Reserves was given by Senator Feeney, Parliamentary Secretary to the Minister for Defence.[655] Feeney had also presented at the 2010 conference. He immediately confirmed that the former 'Rebalancing Army Review Implementation Plan' would not proceed, much to the relief of everyone on the Reserve community. Feeney then provided an overview and confirmed the goals of Plan BEERSHEBA (for which he claimed credit as the instigator), being implemented to restructure the Army Reserve to 'generate integrated capability outputs'. He also claimed credit for Plan SUAKIN, a parallel plan to review the ADF's conditions of service and especially optimise the Reserve's contribution to the Total Force by examining the Reserve employment model 'to better align Reserve capability requirements with employment conditions.'[656]

The complimentary presentation to Feeney's was from Major General J. Caligari, at the time Head, Modernisation and Strategic Planning—Army.[657] Caligari noted that Plan BEERSHEBA still had to be approved by Government in October. He detailed the changes that were coming to the new multi-role Regular brigades as well as the six Reserve brigades. The changes would be far reaching both in the regular force as well as the Reserves, but at least the integrity of the six Reserve brigades remained intact. Caligari concluded:

> This is not a Green paper, where we get another group of experts to go around and interview a whole lot of people and come up with another solution for what the Reserve looks like while at the same time the Regular Army does not essentially change. This is about a complete shuffle of everything.[658]

DRA president Major General Barry acknowledged that Plans BEERSHEBA and SUAKIN provided about 80% of what the DRA was advocating for, especially in regards to a defined role for the Reserve within the Total Force and the resources to carry out those roles, as well as in conditions of service. However, he warned that the DRA needed to concentrate on the 'integrity of the ARes corps elements of armour, artillery, and engineers; ensure that proper technical control and oversight is available for each ARes brigade headquarters and enhance the role of university regiments'. Barry reported that the DRA review panel had a meeting with Caligari scheduled for 23 September with a follow-on meeting with the new Chief of Army, Lieutenant General David L. Morrison AO. Morrison confirmed to Barry that he was 'determined to achieve a balance that allows for significant capability development of our Reserve Force into the future.'[659]

In Victoria, Melbourne University Regiment was celebrating its centenary year. A former cadet graduate of MUR, Major General Greg Garde, later became the last Commandant Colonel at its sister unit, Monash University Regiment (MONUR), where he had earlier in his career been its commanding officer.[660] However, soon after the conference, it was announced that under Plan BEERSHEBA, Monash University Regiment would be disbanded, and its company would be subsumed by MUR. It led to a storm of protest, but to no avail. Colonel James ('Jim') Wood, RFD ED, a former CO of MUR, led the reaction against the decision, calling it 'the demobilisation of ability' in a heartfelt open letter to the Reserve community.[661]

Other speakers at the 2011 DRA conference included Major General Paul L. G. Brereton AM RFD, as Head, CRESD, who reported on the current state of the ADF Reserves. Brereton pointed out that at the time of the conference, almost all of the ADF's contribution to the Regional Assistance Mission to the Solomon Islands and about half of the ADF effort in East Timor was provided by reservists; most of the personnel for the Special Operations Task Group of the winter rotation in Afghanistan came from Reserve commandos and the camp maintenance team in Karin Towt from Reserve engineers:

> Since the Sydney Olympics in 2000, the ADF Reserves have become the default solution for provision of low-risk search and security task groups for domestic security operations and have also become first responders of choice for many disaster responses. Reservists continue to back fill many staff appointments when their permanent incumbents are deployed. Not since 1945 have citizen sailors, soldiers and airmen been so extensively committed to current operations.[662]

Following Brereton, Major General Irving, DRA vice president, followed with a paper 'Using the ARes to Enhance Army Capability'. He clearly enunciated his and the DRA's, view that the:

> paradigm shift [over the past decade in Reserve deployments] and consistent delivery of capability have demonstrated that properly structured and resourced the ARes can be trusted to deliver on capability, either in a shared role with the ARA or as a repository for longer lead time and niche capabilities. It has and can embrace change.[663]

In keeping with the DRA's policy to ask both Government and Opposition to the conferences, the Opposition view on ADF Reserves capability was provided by Stuart R. Robert MHR (Liberal/Fadden), Shadow Minister for Defence Science Technology and Personnel.[664]

Two other important presentations were made, taking up one of the conference goals to focus on the Navy Reserve. 'The Navy Reserve in the Future Navy' was addressed by Commodore R. Phillips, Director General Reserves—Navy.[665] Phillips noted the two elements comprising the Naval Reserve, namely the Complementary Reserve which provided specialist skills and the Supplementary Reserve which comprises primarily ex-Permanent Navy members. The DRA's vice president Navy, Captain J. Lukaitis, also spoke to the topic of the role of the Naval Reserves into the future.

Lukaitis reflected the DRA's long held view that cuts to reserve numbers and training days were done for financial reasons rather than any strategic plan rationale—citing the cut back in Active Navy Reserves working more than 20 days a year being reduced by 25% over 2010/2011, from 1577 to 1165 reservists. He also decried 'the current level of negligible engagement and communication with non-working reservists'. The recent retirement of Lukaitis, followed by Commodore Phillips soon after, saw the last of the 'career reservists' in the Navy Reserve. Career reservists at seaman and senior sailor levels had almost disappeared as well by that time.[666]

Lukaitis was scathing when it came to current policy towards Naval Reserves: 'To continue using the Naval Reserve as an *ad hoc* mechanism for filling gaps in Navy capability represents a lack of imagination and foresight, poor personnel management and may even prejudice Navy's ability to meet future operational requirements', but he did offer up a range of practical and cost effective solutions to the issues at hand.[667] Major General Barry, in a later issue of *The Australian Reservist*, added his observations on the situation in Navy.

Barry noted that the Navy was struggling with Reserves for there was very little recruitment off the street and on the other hand little incentive to stay in the Reserve given that out of about 9000 reservists in the Active and Standby Naval Reserve (most ex-permanents, only 1165 found work for more than 20 days a year—'The Naval Reserve is currently a wasted asset and badly needs a purpose.'[668] After the conference, Senator Feeney was presented with the expanded copy of the paper presented by Captain Lukaitis at that conference 'The Role of the RANR—2011 and the Future'. In the months after Feeney sent the paper to be 'staffed' first by his own office and then by Navy, and he returned to it at the conference in 2012.

DRA president Major General Barry reported to the DRA AGM following the 2011 conference that:

> Since our last conference [in 2010] it feels like we have been marking time and not much has been achieved and that is a reasonable viewpoint to reach. However, much work and advocacy has occurred since the Government appointed four new Defence Ministers and secretaries on 12 September...[669]

However, he also noted that there had also been changes to the ADF hierarchy in July and 'we have to understand if there are any further shifts in direction...Only time will tell'. The AGM saw changes to the Executive. Captain P. Wertheimer stepped down as national secretary, being replaced by Major Bedggood. National treasurer Colonel Dennis Townsend also stepped down to be replaced by Lieutenant Colonel Jennifer ('Jenny') Cotton.[670] The issues had changed little from the previous year and branches had little new to report; there were no reports at all from Tasmania or the Northern Territory.

Private Luke Goodwin (R) points out the arcs of fire to Lance Corporal Alex Tyler on Exercise ANZAC, part of the Junior Leadership Course held in Timor Leste in 2011 for ARes soldiers from the International Stabilisation Force.
(Defence Images)

On 23 November 2011, DRA president Major General Barry met again with Major General Caligari. The meeting 'confirmed the majority of the review panels opinions and recommendations around Plan BEERSHEBA, with the exception of the final structure for Reserve combat arms', had been accepted. Plan BEERSHEBA was put to the Government and formally launched in December 2011 by the Minister for Defence Steven F. Smith (Labor/Perth) and CA, Lieutenant General Morrison. [671] Barry was invited to the East Coast launch led by Senator Feeney.

While Plan BEERSHEBA would take years to roll out, nonetheless, in the words of Barry, '...the Army Reserve after 60 years is moving from a mobilisation/expansion base of 'just in case', to an operational role of 'just in time, with a clearly defined role and tasks'.[672] It was perhaps, a surfeit of enthusiasm over experience. As the Government's budget of May 2012 was to show, the Reserve still had many trials ahead. In the meantime, the DRA's objective is to improve implementation of the plan, rather than be a thorn in the side.'[673] Nonetheless, Major General Caligari wrote to the DRA president following their meetings in September and November 2011 and expressed his thanks 'for the constructive approach your association has taken to Plan BEERSHEBA as it made a very useful contribution to not only BEERSHEBA but the progression of the Total Force Concept.'[674]

Into 2012, the DRA continued to work in the 'human resources' areas of the three services and their Reserves, within Plan SUAKIN, which like Plan BEERSHEBA, had its origins in the DWP'09. CRESD had already been working on elements such as the Civilian Skills database, the Reservists Personnel Cost Model, and a predictive Behaviour Model. With the new consultative environment arising from Plan BEERSHEBA, the DRA now had the opportunity to also get involved in Plan SUAKIN as well, being able to add opinion and recommendations to the streams of service categories (SERCAT) with its addition of 'non-permanent part-time' and 'permanent part-time' as well as to Service Options (SEROPS) with its addition of 'project work' and 'dual employment'. DRA members, especially those retired officers who had left the service some time ago, would have felt very dated by this rapid pace of change, not just to the structure but detail behind it.

On the surface at least, and unlike the past reviews and restructuring, it did not seem that the Reserves would suffer any more unit disbandment or dislocation, but some impacts would be inevitable as the Reserves changed posture to meet the new brigade capability imperatives. However, Major General Jeffery J. Sengelmann DSC AM CSC, successor to Major General Caligari as Head Modernisation and Strategic Planning—Army, wrote to DRA president Major General Barry in April 2018 to confirm that planned reductions in unit sizes in the artillery, construction regiments and Monash University Regiment would go ahead despite the DRA reservations about these aspects of Plan BEERSHEBA. Nonetheless, Sengelman welcomed the ongoing dialogue and consultation with the DRA and offered updates and briefings on the implementation of the Reserve restructure.[675]

The 2012 DRA conference was well into its planning phase by April 2012—with the title 'ADF Reserves—Maintaining the Momentum in Testing Times', a prescient title in view of coming changes. In line with the promise that the RAAF Reserve would be one of the *foci*, the DRA had established a working party to examine the Air Force Reserve in detail with a view to reporting to the August conference. Major General Barry noted that while the Air Force Reserve, a totally integrated workforce like Navy's, appeared to be in better shape than Navy due to its retention of City Squadrons and their merger with Base Squadrons. At the same time, and like Navy, it had about 9000 reservists. However, in a similar case to Navy, only 2080 personnel had worked more than 20 hours in 2011. 'There is', said Barry... disquiet on the people front as they are not being well managed and are expected to respond like their permanent colleagues.'[676]

In May 2012 came the Labor Government's budget for 2012/2013. Its main goal was political, that is to achieve a budget surplus and demonstrate that Labor was as effective as the

Coalition opposition in terms of economic management. More than $4 billion of spending on defence projects was reduced or deferred. While there were few retired reservists who would have objected to almost 1000 civilian positions being eliminated within Defence, the Reserves were hit hard by a 10% across the board reduction in budget. The DRA president drew upon an ASPI review of the Defence budget to explain to DRA members the impact.

According to ASPI, according to a survey at the time, only 1% of the Australian public thought national security was important. The Government no doubt had done their own surveys too. According to the ASPI critique, defence spending as a percentage of GDP was the lowest since 1937 and would go even lower in 2013/14. Personnel costs were reported to be 43% of the Defence expenditure. Barry said: 'With the cost of personnel at that level it is easy to see why the Reserves take a hit and it is now all too obvious.'[677]

Barry explained the ramifications for the Army Reserve while noting that Navy and Air Force Reserves were not affected by the cuts as they had already pared down their use of Reserves over the last two to three years, so their salaries would be about the same. For the ARes, Barry cited the example of a Reserve Battalion Battle Group under Plan BEERSHEBA which would need 50 days pre-deployment readiness training. Under the cuts this meant that other reservists would not get the generally mandated 20 days of training each year. That in turn would result in them being non-efficient in terms of long service awards, health support and housing loans. This when it was recognised that an infantryman needed 30-37 days to reach a basic level of effectiveness, outside of training courses.

'The effect of these cuts', said Barry, '...will have a major morale impact on reservists in them not feeling they are required, or worse still, wanted'. Barry anticipated that many would vote with their feet and leave, especially as only 25% would now be able to access the 'entitlements.'[678] The DRA began to lobby for commanding officers to deem 'efficient' soldiers who had not been able to complete the mandated days through no fault of their own (the authority to do this had to be devolved to COs). However, the entitlement question was another issue altogether. Barry recognised that the Association had been in this position before:

> We the Reserve Forces, have been through this this situation a number of times before with predictable outcomes, but we do not seem to have learned the lessons of history. The Services must communicate with their reservists and tell them the facts if they want to keep them. "Don't shoot the messenger", as it is the Government which made the cuts! ...We have been there before, and we will weather the storm with a little help from our friends.[679]

The cuts affected the Reserve in many ways. In Tasmania for example, the cuts meant that 16 Battery RAA, which had converted to mortars from guns with minimum disruption, was to be disbanded, destroying a lineage of 150 years in the making. The battery was to be amalgamated with another in South Australia to form a new composite battery with a strength of 68 (compared to 16 Battery's 80). Colonel Stephen ('Steve') Carey, DRA president in Tasmania and Honorary Colonel of RAA units wrote to the Minister for Defence objecting to the plan. He considered that the composite battery plan 'was set up to fail' and would result in 'significant community dissatisfaction'.[680]

With cuts announced the DRA conference for 2012 went ahead soon after on 28 July 2012, again in Keswick Barracks in Adelaide; about 80 attended. The Plan BEERSHEBA update was provided by the new Head Modernisation and Strategic Planning—Army, Major General Jeffrey ('Jeff') J. Sengelman DSC AM CSC. Sengelman's paper emphasised the 'total integration of effects' and outlined the first structural changes affecting the paired ARes brigades to provide an ARes Battalion Battle Group by the end of 2013. Many DRA members would have worked hard to keep up with the rapidly evolving terminology of the modernising Army and Reserves—from the Hardened and Networked Army to Enhanced Land Force, the Adaptive Army Initiative to the Force Generation Cycle.[681] Additional Bushmaster protection mobility vehicles (214) had been ordered as part of the modernisation plans but the last reserve brigades would not be equipped until 2017. Sengelman particularly noted the DRA's contribution towards these changes especially with regard to Plan SUAKIN, where the DRA's review panel had been working with Major General Brereton, CRESD.

The DRA had commissioned, leading up to the conference, an ASPI critique of Plan SUAKIN, but as Major General Barry pointed out, the critique—presented by Dr H. Smith, a Visiting Fellow at ADFA—'actually became more of an endorsement'. Smith noted that the Millar Report into the CMF in 1974 had actually put forward a number of recommendations which were finally, under Plan SUAKIN, coming to fruition more than 40 years later. It was a time cycle for progress with which the DRA was depressingly familiar! However, Smith said that 'Plan SUAKIN was necessary whatever the role of Reserves'...and that it 'provides a road map and a route to the promised land where the very word 'Reservist' might be abolished or simply fade away'.[682] Dr Samantha Crompvoets, a Research Fellow at the ANU School of Medicine, also spoke to the conference about the culture and change management aspects of Plan SUAKIN.

Air Commodore Delahunty, Director General Reserves—Air Force, spoke at length on developments in the Air Force Reserve. The Chief of Air Force had issued a strategic intent update in February 2012. It rested on the Air Force experience with integrating their permanent and Reserve workforce since 2005 and formalised the change of focus to support of operations rather than the traditional concept of Reserves as a having strategic surge capacity. Delahunty noted that even in 2012, there were still those in both Reserves and permanent elements who held 'pre-integration attitudes'. One new role for reservists mentioned was as operators of remote piloted vehicles.[683] Brigadier Spence, the Director-General Reserves-Army, reported some disturbing figures: 58% of ORs and 65% of officer trainees separate before being qualified.[684]

Senator Feeney, Parliamentary Secretary to the Minister for Defence, returned to the DRA conference to provide the Government's viewpoint. The DRA had developed a good relationship with Feeney and Feeney had reciprocated. Feeney was seen as sincere, interested, and was well regarded. But Feeney pushed back on two issues. Senator Feeney's Defence Determination 2012/68 of 14 December 2012 followed by ill-researched and negative comments in Parliament on 14 May 2013 regarding self-employed reservists potentially rorting the Employer Support Payment Scheme (ESPS) had led many in the Reserve community to describe the Senator in less than complimentary terms. At the conference he related that 43% of ESPS payments were going to self-employed reservists, sometimes inappropriately.[685] Some reservists argued that Parliamentarians who were paid to be MPs and then received remuneration from Committee work in addition were no different.

Senator Feeney, in a follow-up to the Lukaitis paper from the year before noted that discussion and analysis was already underway in the lead up to the new Government DWP 2013. However, based on feedback from Navy (which later proved to be inaccurate), he pushed back against some of the recommendations in the Lukaitis paper. Lukaitis, a highly experienced and senior Naval Reserve officer who had worked for years in the senior planning areas of Navy, in turn pushed back against some of Navy's assumptions. The result was, with Feeney's support, the announcement some months after the conference of a new Navy inquiry into the state of the Reserves under new Chief of Navy, Rear Admiral Raymond ('Ray') J. Griggs AO CSC.[686] Lukaitis and the DRA continued their efforts to highlight deficiencies in the Navy Reserve system.

Following Feeney was once again Stuart Robert, Shadow Minister for Defence Science, Technology and Personnel. His point that there had been three Defence Ministers in five years was a strong one, and underlined the DRA's position that two year postings for Reserve officers and permanent officers alike were too short, not to mention the turnover among politicians associated with Defence as well. Indeed, presentations at the DRA conferences over the years by revolving holders of the same position highlighted this very point. Regarding the budget cuts, Robert claimed that 28 days out from the budget the CDF and Service Chiefs had not been informed that the cuts were coming. He pointed out the hypocrisy in the decision to announce issue of Bushmasters to Reserve units while at the same time in effect reducing the training hours available to bring their crews up to standard to use them.[687]

At the DRA AGM on the day following the conference, various national issues were discussed. Among them were superannuation; DRA policy review (underway through DRA vice president Major General Irving but awaiting implementation of Plan SUAKIN); the ongoing and successful Tasman Scheme selection process; downsizing/disestablishment of university regiments; cancellation of the 'gap year' program for tertiary students; status of Service bands under threat; and the planned Operational Service and National Emergency Medal.

Continued pressure regarding declining membership was reflected in ongoing encouragement from the DRA president for branches to keep working on succession plans to help reinvigorate committees.[688] This was not new. Attracting Active reservists as members was challenging when so many were on full-time service and there were fewer and fewer Reserve officers in senior roles in the States; many of those roles which had existed previously were now national roles in Canberra or filled by serving permanent officers.

Branch reports gave their annual updates. Queensland branch reported only 30 ordinary members but continued with information sessions for members on different topics of interest attracting between 40 and 55 members and guests. NSW branch reported a good number of members—135 ordinary and 101 fully subscribed—and was busy with numerous events and activities. Victoria reported 149 ordinary and 93 life members. It boasted more than 20 on its committee, several of whom also directly supported the DRA president and National Executive. Victorian president Major General Garde was replaced by Brigadier Alkemade, a former Commander of 4 Brigade. Garde had been appointed a Supreme Court judge and president of the Victorian Civil and Administrative Tribunal by the State Government; those appointments also put paid to plans afoot to elect Garde as DRA president to replace Major General Barry.[689]

Tasmania also saw a change of guard with Lieutenant Colonel Townsend the former president retired from committee along with another former president, Lieutenant Colonel Douglas ('Doug') M. Wyatt, OAM RFD. In stark contrast with the situation two years later, Tasmania reported an apparently robust 150 life members and 60 ordinary members, but also admitted that 'we have a significant task ahead to revitalise the number of members to 'paid up' status and to attract serving members ....without this we will be obliged to use invested funds to meet operational expenses.'[690] South Australia, under its president, Major General N. Wilson, noted a good increase in member numbers to 87 using the device of a complimentary membership in the first year. However, drawing in serving members 'remained a struggle'. Western Australia reported 148 members but only 11 were serving.[691] While membership numbers appeared healthy, in fact the ageing demographic and the lack of success in attracting serving members continued to be of concern.

Leading up to the annual conference in August 2013, by April the DRA had commissioned a Strategic Overview of the Indian Ocean together with an examination of border protection issues to the near north of Australia. In particular, the review was to examine the role of Regional Force Surveillance Units (RFSU).[692] The DRA had also requested updates on the two major structural programs underway in the ADF namely Plans BEERSHEBA and SUAKIN. Unresolved, more detailed issues were recognised as 'grist to the DRA mill'. For example, Major General Garde's amendments to the *Defence Reserve Service (Protection) Act 2001* had passed the scrutiny of the three Services, were regarded as cost neutral and beneficial and received bipartisan political support. However, they were withdrawn from Parliament's legislative agenda. As DRA president Major General Barry was to describe it in *The Australian Reservist*, 'in fact they have been sent off in classic fashion to be considered within the context of Plan SUAKIN...even though the Protection Act has no direct relevance to contemporary work place reform, either in remuneration or structure.'[693]

Another issue was inherent to the Defence Home Ownership Assistance Scheme Act 2008 (DHOAS). An amendment to the Act would disadvantage reservists, calling for eight consecutive years of effective service to qualify, rather than cumulative years. Barry again: 'This amendment is seen as a deliberate attempt by the Government to save money and runs counter to the scheme's intention which is to aid retention of personnel in the Reserve.'[694] With regard to the ESPS, Determination 2012/68 by the Parliamentary Secretary for Defence, Senator Feeney, which came into effect in January 2013 upset many.

The Determination reflected Feeney's concern that Self Employed Reservists (SER) would double-dip in the scheme and so their Reserve employment conditions were made especially restrictive, almost to the point of discriminating against them. Barry observed that historically more than one third of reservists eligible for payment under the scheme were SERs. He also noted that the Defence bureaucracy had managed to set approvals for applications under the scheme, even for 'other ranks', at one-star level. This 'bureaucratic complication' added another barrier to SERs giving effective service.[695]

The restriction to Service Days, a result of the budget cuts in May 2012, also continued to be a major issue. Many personnel were missing out on being classed as efficient and so become ineligible for time in rank, long service awards, health support allowances and the DHOAS loan. DRA maintained that COs should be able to certify efficiency for at least time in rank and LS awards, a change that could be directed by Service Chiefs. The DRA was also concerned about the base/depot review. It felt that that each ARes unit in the ORBAT should submit a long term plan for facilities to ensure their viability and link with other service organisations to achieve economies of scale.[696] The 2013 Defence White Paper was released in early May 2013. There had been no community consultation process, 'it was not tabled in Parliament, and it lacked bipartisan support.'[697]

On 14 August 2013 Major General Barry released his bi-annual Executive Circular—his 25th and final as DRA president. He listed the many activities in which he had participated as president since the start of the year. It showed that he had been involved in some 50 appointments, briefings, meetings and events. Considering the voluntary nature of the position, Barry had maintained an energetic program which involved not only maintaining and nurturing the Association itself in all of its internal organisational and administrative development, but also helping to keep Reserve issues and the DRA itself in consideration of politicians and Defence officials and senior officers alike.

By way of comparison, his first Executive Circular in October 2005 noted that he had joined in just three events; two years later this had risen to over 21 and by 2011 to over 40 in a six-month period. There is no doubt that when added to the many hours of work in telephone conferences and calls, emails, and general correspondence the workload was indeed a heavy burden. Regardless of who was in the role, this workload was reflective also of the many others in the executives at both national and State level who worked tirelessly for the Association over the many years of its history.[698]

The DRA conference on 17 August 2013 was held for the first time in Western Australia and for the first time at a Navy establishment, Fleet Base West, HMAS *Stirling*. The theme was 'Developing Reserves Capability in Challenging Times'. It was also held just four weeks before the federal election, which was to see the Labor Government of PM Kevin Rudd swept from office and replaced by the Coalition under Anthony ('Tony') J. Abbott MHR (Liberal/Waringah). Nonetheless, both former Senator Feeney and Stuart Robert MHR attended.

Updates were heard on Plans BEERSHEBA and SUAKIN, and the overview of the Indian Ocean and RFSU (in this case NORFORCE) were presented, along with the traditional Service 'report cards' on their Reserves. Major General Spence, at the time of the conference

Director of Project SUAKIN, presented on the 'SUAKIN Total Work Force Employment Model', noting that although Defence had aspired to this for some time, the Services were 'at different levels of maturity'. Hence SUAKIN, designed to create a comprehensive, coherent and consistent workforce system.[699]

At the DRA AGM that followed the conference, Major General Barry stepped down and was replaced by Major General Irving. Major General Flawith moved to Deputy DRA president and Major General N. Wilson as Vice-President Army. Commodore Elsey stepped down as Vice-President Navy and was replaced by Captain Joe Lukaitis.[700] Barry had made a number of significant changes to the way the DRA operated during his tenure as DRA president. Noting the Army and 'east coast' orientation of the DRA when he succeeded Major General Glenny, Barry introduced vice-president positions for all three services and a vice-president position for Defence Health, roles also introduced in the State branches as well.

Barry had also begun the process of taking the DRA nationally out to the States through carefully prepared conferences run to set themes (a concept started by Barry's predecessor, Major General Glenny). Conferences were subsequently held in traditional venues like Randwick Barracks in Sydney, but also at RAAF Williams in Victoria, at Keswick Barracks in Adelaide as well as in HMAS *Stirling* on Garden Island near Perth. He had also successfully used a small Victorian task force and 'think tank' to assist him as national president and gained approval from Army to form a review panel of retired Reserve 'two-stars' for liaison with Army especially around Plans BEERSHEBA and SUAKIN.

The following AGM meeting discussed a wide range of issues both internal and national. Among other matters discussed were the need to update DRA policies to align with Plan SUAKIN and the rejuvenation of the DRA website. Like other volunteer organisations engaging with the internet, the DRA had discovered that maintaining a website and keeping it relevant and contemporary needed ongoing time, money, and expertise. In another chip off tradition, the meeting heard that Honorary Colonel positions had been abolished at university regiments.[701] The DRA called this a 'short-sighted, no cost saving, mistake!'. It noted that these appointments liaise with vice-chancellors and opinion makers for the public relations benefit of Defence, and called for a review of that decision.[702]

In a reminder that the DRA could actually play a more direct role in supporting the Reserves, the Victorian branch reported that it continued to provide direct support to reservists and their families on and after operational service. It had done this during Operations ANODE and ASTUTE (East Timor), most recently for the Regional Assistance Mission to

Solomon Islands (RAMSI) as well as in East Timor, not to mention support for individuals deployed on Operation SLIPPER in Afghanistan.[703]

In late 2013, the new Abbott government saw a promise to release the next Defence White Paper within 18 months, 'costed and affordable'. The Government aimed to achieve this by ensuring there were no further cuts to the defence budget and that they would work towards raising defence spending to two per cent of GDP. Both the 2014–15 and 2015–16 budgets significantly increased spending on defence.[704]

In July 2014 the Minister for Defence Kevin J. Andrews MHR (Liberal/Menzies) released a discussion paper on the development of DWP2015, which 'raised the important question of what the optimum use of reserve forces should be in enhancing ADF capability'. The discussion paper pointed out that the DWP2015 needed to define the most appropriate roles and missions for Reserves and that they be achievable, given the level of Reserve training and equipment. The DRA, according to DRA president Major General Irving, was heartened by these issues being raised and created a small writing team was formed to prepare a DRA submission to the DWP2015 process—that submission was due on 29 October 2014.[705]

April 2014 saw the first message in *The Australian Reservist* from new DRA president Irving. There was much to report, starting with an imminent decision on the repeal of the sunset clause in the regulation governing the ESPS. The Service chiefs had agreed to fund

Army reservists from the 12th/40th Battalion, The Royal Tasmania Regiment, training in waterborne patrolling with a Zodiac from the Hobart-based Navy Reserve Diving Team 10, in 2014.
(Defence Images)

the ESPS—'there is no doubt that the ESP provides considerable capability to the ADF at a moderate cost'.[706]

However, no outcome was in sight with regard to the DHOAS as it applied to reservists, despite DRA lobbying to interpret the legislation for cumulative years of service rather than consecutive. Irving also reported the hope that proposed amendments to the *Defence Reserves Service (Protection) Act 2001* would be considered by Parliament in the forthcoming winter session. The amendments included removal of references to 'unprotected service', stronger anti-discrimination provisions, education to be protected and civil penalties introduced. This hope in the context that the Garde review of the legislation was completed in 2008.[707]

Irving reported on progress with Plan SUAKIN. He reported that Phase 1 was expected to be completed by June 2014 with the responsibility for implementation transitioning from CRESD to the Defence People Group. Funding for Phase 2 had been approved by the Secretary Department of Defence and the CDF Advisory Council. Irving promised that the DRA would watch implementation closely given that Plan SUAKIN 'was essential in developing a Total Force concept where there is a mix of employment categories working together to provide capability to the ADF'. Another significant development in the concept was that the CDF no longer needed Department of Finance approval to deploy reservists, another falderal put to rest on the pathway to Total Force becoming a practical reality. Meanwhile the national DRSC continued to initiate strategies to improve and enhance the employer support for reservists. Irving highlighted one initiative, a dual employment model where a formal agreement is entered into between Defence, the reservist, and the employer to 'share' the reservist.[708]

Plans were already underway for the annual DRA conference, to be held in Canberra on 23 August 2014 with the theme, 'ADF Capability through Flexibility'. The theme aimed to illuminate the flexibility afforded the ADF by Reserves at a time of budget pressures. Remarkably, within DRA conference history, former Senator and now MHR, David Feeney (Labor/Batman) and now Shadow Assistant Minister for Defence, attended for his fifth conference in succession. Stuart Robert, the Assistant Minister for Defence was absent on Government business overseas but was represented by newly elected Senator Zdenko ('Zed') M. Seselja (Liberal/ACT). The VCDF, Vice Admiral Ray Griggs, also attended part of the conference.

Conference speakers emphasised the importance for Plan SUAKIN to be fully implemented, to achieve a modern and flexible work force. Other issues affecting Reserves were also to the forefront at the conference, for example, the issue of reservist service and the DHOAS. The DRA's 'cumulative' years of service focus was subsequently raised with Assistant Minister Robert's office, but by October, no response had been received.[709] Other speakers included Dr A. Davies, Senior Analyst for Defence Capability at ASPI, who examined the strategic fundamentals underpinning the next DWP2015. Commenting on previous White Papers, while cautioning against the inexperience behind another first year in office DWP2015 he said:

> ...the 2009 Defence White Paper can now be regarded mostly as an historical curiosity. And the 2013 White Paper was such an obviously hollow political exercise in covering for a shameful pillaging of the Defence budget that we shouldn't concern ourselves with analysing it. In fact, starting from a much lower budgetary base, it nonetheless increased the capability aims from its 2009 predecessor, with no semblance of a funding plan.[710]

At the conference, Major General Sengelmann, Head Modernisation and Strategic Planning—Army, presented on the Army's modernisation initiatives. One of the recommendations moving forward was to disband—or 'dis-establish'—the HRR in October 2014. Following the presentation in 2013, the report to the 2014 conference by Major General Steven L. Smith AM CSC RFD on 2 Division's progress was notable in this respect, especially as he concluded by noting the challenge posed by meeting the competing priorities that the Division faced. 'Any one of these issues in isolation is a challenge in itself', he said, 'Compounded, they represent a complex series of issues to be navigated through'.[711] Major General Spence, CRESD, returned to the conference gave an update of key programmes impacting reservists.

At the follow-on DRA AGM on 24 August 2014, Group Captain Schiller was elected Vice-President Air Force. That AGM saw a big focus on membership. State branch membership continued to change as they had over all the years of the Association's existence; some States had developed succession plans to 'reinvigorate their executives to ensure their committees remain active and relevant.'[712] SA branch president Major General Wilson noted a stable membership of around 79 members but also that there appeared to be little enthusiasm to revive Reserve Forces Day, which had not been conducted for the past three years. Like other branches it maintained its representation on the State DRSC and cooperation with other ESOs.[713]

Colonel Carey reported from Tasmania failure in even holding committee meetings due to a lack of a meaningful quorum, while 'Ongoing endeavours to attract ordinary members have met with no positive results with recently retired reservists "wanting a break" and active reservists "too busy".'[714] By contrast, the WA branch under Lieutenant Colonel Richard ('Dick') G Cook AM RFD reported 151 members (but only including 15 serving reservists and only one RANR member).[715] Lieutenant Colonel J. Cotton reported from Queensland a small membership of only 20; it was represented on the State DRSC as well as the Queensland Consultative Forum hosted by the DVA (Qld.).

DRA president, Major General Irving, reported in October that the proposed amendments for the *Defence Reserve Service (Protection) Act 2001* now appeared to be scheduled for Parliamentary discussion in early 2015, seven years after the amendments had first been proposed; at this point it was not even clear, according to Irving, whether there would be bi-partisan support for them, which was a rather startling admission this close to a possible denouement for this long standing saga. Irving could not help but wonder: 'How long would it take to progress the legislative amendments to implement Plan SUAKIN?'.[716]

Nonetheless, in many areas, the Army's modernisation programs were progressing and by 2025 it was thought that the Total Force (Permanent and Reserve (or P/T) components would see an authorised level approaching 50,000. This proved to be an underestimate of progress. By 2020 there were about 79,700 in the ADF, comprising 60,000 Permanent and 19,700 Reserves.[717] Yet in 2014, for all the progress, or perhaps because of it, a major gap came to light in that little effort had been made to manage Standby Reserves. It was revealed that the services did not even have an up to date contact list or really much idea about their availability.

In March 2015 the DRA president Major General Irving, met with the Assistant Minister for Defence Stuart and his staff to discuss the range of issues on the DRA agenda for resolution. The DRA submission to the DWP2015 was also discussed. Issues included enhancing the role of university regiments, the alarming state of Reserve recruiting, superannuation for reservists, eligibility of reservists for DHOAS, proposed amendments to the Protection Act (again) and implementation of Plan SUAKIN. By February 2015, Reserve numbers had dropped to 13,700—more than 3000 less than five years earlier.[718] Regarding university regiments, the DRA advocated that the units should be able to recruit across all university campuses and recruit for both F/T and P/T elements of the ADF. Irving put it to Assistant Minister for Defence S. Robert that the centralised model for new entry (or *ab initio*) recruiting was not working and that Reserve units and formations should be properly resourced to do their own. This was not a new idea but a return to a model which had been proven to work in the past, and for Reserves, supported their 'from the community, of the community' ethos.

Irving reported in *The Australian Reservist* that the Minister supported in principle superannuation for reservists but linked it to taxation of Reservist pay—the DRA rejected this linkage. Irving's response was automatic; the DRA had fought taxation of Reserve pay 20 years ago. There was no appetite for a discussion on this especially when 365 days was still the ruler for determining daily pay rates. In the meantime, Defence had rejected the DRA view that reservist eligibility for DHOAS should be based on cumulative years of service; it claimed it would cost an extra $11 million based on 2600 reservists becoming eligible each year. That issue remined unresolved and the parliamentary review of amendments for the Protection Act was as once again postponed to late 2015.[719]

Plan SUAKIN was progressing albeit with some 'tweaks' to the conditions of service in some SERCATs. At the meeting, the Minister advised that it would be unlikely that DWP2015 would be released before the DRA conference planned for 22 August 2015 in Sydney.[720] As part of the DWP2015 process, the Government has been conducting a community consultation process. The final report on the outcomes of the process around the release of the Issues Paper in July 2014 was announced on 1 July 2015.[721]

In August 2014 the previous Minister for Defence had commissioned a First Principles Review of Defence; the findings were released by the new Minister for Defence, K. Andrews on 1 April 2015. The purpose of the review was 'to ensure that [the Defence Department] is fit for purpose and is able to deliver against its strategy with the minimum resources necessary'. It delivered 70 recommendations, among them: 'Defence to define the estate need as determined by future force requirements and Government agree to dispose of all unnecessary estate holdings starting with the 17 bases identified in the 2012 Future Defence Estate Report.'[722]

At a later DRA conference, speaker David Feeney MHR explained the situation:

> The Army Reserve must confront the challenge of a withering and poorly situated estate. The legacy estate, largely from 1945, presents at least two serious deficiencies. First it isn't where we need it. Large swathes of metropolitan and regional Australia have no Army Reserve presence because they have arisen since 1945. Secondly, the estate we do possess is old, often poorly maintained and often under-utilised...With 235 facilities across Australia valued at $1.7 Billion, together we should explore new modes of organisation.[723]

The DRA immediately had concerned correspondence especially from South Australia regarding the possible closure of Reserve Depots there, and elsewhere. It was not the first time this had happened either, but the ARes Association had been powerless to stop closures then. Underlying any appeals on historical or traditional grounds not to close depots the very real concern based on bitter experience that closing depots also meant possible amalgamation or even disbandment of Reserve units.

The 2015 DRA conference theme was 'ADF Total Force—Fact or Fiction'. The conference aimed to examine the capability provided by reservists from each service to the total ADF. Due to increased security at ADF bases and the difficulty in finding the facilities and accommodation required on a base for the second year running the conference was forced to take up a hotel setting, this year in Sydney. It was a far cry from the earlier and more engaging conference environments associated with Service base and mess facilities, but only about 70 attended the conference despite efforts to advertise and contact members through emails and letters.[724] The now Governor of NSW, General Hurley, opened the conference. Keynote speaker was Major General John Crackett CB, UK Assistant CDF—Reserve Forces and Cadets who presented on the UK experience of implementing their 2013 White Paper.

Representing the Government view was Senator David J. Fawcett (Liberal/SA), a former Regular officer and a future Assistant Minister for Defence. He attended the conference as Chairman of the Defence Sub-Committee of the JSCFADT. He made the observation that 'the ADF Total Force is an evolving reality with reservists playing an increasing important ADF capability locally, nationally and on overseas deployments'. His counterpart at the conference, the experienced David Feeney, at his sixth DRA conference, made a similar observation: 'The evolution of Army Reserve towards a force able to provide military response options to short term contingencies (rather than a militia designed for mass mobilisation) must continue.'[725]

Other speakers included Rear Admiral Jennifer ('Jenny') R. Firman, Surgeon General ADF—Reserves, who emphasised the need for better tri-service integrated planning for defence medical specialists. Rear Admiral Michael J. Slattery AM RANR, Judge Advocate General, provided an informative presentation on the considerable input to ADF capability by Reserve legal officers while Major General Fergus ('Gus') McLachlan AM, Head Modernisation and Strategic Planning—Army (HMSP-A) spoke about current and future proposed initiatives to enhance the ARes. He also emphasised that new ARes facilities should be leased or built before old facilities are closed.[726] Major General Spence, CRESD, once again returned to the conference to give an update of key programmes impacting reservists, and noted, to nobody's surprise that the amendments to the Protection Act had once again been postponed in Parliament until 2016.[727]

Commodore Bruce Kafer AM CSC, DGANCR, advised of a 'roots and branch' review of the RANR initiated by the Deputy Chief of Navy. The review, to be completed by June 2016 would examine the role, structure, size and management of the RANR. This announcement was a major source of satisfaction for the DRA and especially Captain Lukaitis, the DRA vice-president Navy. Their advocacy and promotion of informed debate for changes to the Navy Reserve system, which had begun at the 2011 DRA conference and continued since, heavily influenced the call for a comprehensive review. Naturally, the DRA intended to make a full submission to that Review.

On 27 August 2015, DRA president Major General Irving and Major Generals Barry, Garde and Glenny went to Canberra for a briefing by Chief of Army on the 'expected significant challenges' facing Army with the implementation of DWP2015 when it is released. Invited to the briefing were former two stars and Army RSMs. The participants were shown a 'significant amount of new equipment, vehicles, weapons, and battle management systems to be rolled out in coming years. The question was, would the Reserve be able to access this when they needed it? Shortly after, in a cabinet shuffle in September 2015, Senator M. Payne became the new Minister for Defence and Darren J. Chester MHR (National/Gippsland) the former Parliamentary Secretary for Defence, became the new Assistant Minister. As Irving noted, Senator Payne was the ninth Minister for Defence since January 2011 and only one had served more than four years. A similar situation applied to the Assistant role—'Regrettably, both sides of politics have not given sufficient weight to the importance of these two most important portfolios.'[728]

12.
Disappointment and Optimism in equal measure

It seemed almost inevitable that after what appeared to be an optimistic start in engaging with the DRA over, first, Plan BEERSHEBA and especially Plan SUAKIN and then, in regard to consultations and submissions for what became DWP2016, the DRA president would be downhearted. In April he said that he was 'extremely disappointed that when the DWP was finally released in February 2016, the importance of these submissions had been completely ignored and DWP2016 made no recommendations about the use of Reserve forces' (in fact DWP2016 included only three paragraphs referring to Reserves). Major General Irving generously added: 'While we may never know why the people who were tasked with preparing the DWP ignored the submissions on the use of Reserves, I suspect that this outcome was caused by neglect over the long gestation period involved in preparation of the DWP, rather than the Services including presentation of the Reserves in every capability decision'.[729]

Irving noted in his president's report in April 2016 that other issues remained unresolved. For example, the amendments to the *Defence Reserves Service (Protection) Act* had still not been considered by Parliament. It was an indictment of both political parties, wrote Irving, and 'another indication of the lack of priority given by Government and Opposition to Reserve issues'. Irving did report on DRA initiatives in other areas. For example, at the conferences in 2014 and 2015, the role, training, and platforms of Armoured Corps (RAAC) reservists had been raised. An options paper in consultation with COs was compiled by Colonel Lee T. Long RFD, Honorary Colonel of the 1/15 Royal NSW Lancers. Major General Irving raised the paper with Major General Stephen H. Porter AM (later AO), Commander 2 Division. Porter in turn reported to CA's Advisory Committee.

The Long paper aimed at 'enhancing the role of RAAC reservists and to address the wide divergence that exists between RAAC Regular and Reserve components', and 'emphasised that RAAC reservists must be trained in core mounted and dismounted RAAC skills'. Irving subsequently reported that 'there appears to be acceptance that we need to train RAAC reservists on appropriate platforms if we are to cover capability gaps in the Regular Army.'[730] The question of course, was whether the Reserves would actually be issued with the 'appropriate platforms' or be allowed to train on them in any meaningful sense of the word. The Lancers' regimental newsletter later said, on the occasion of Long's retirement as Honorary Colonel:

> We can credit...Colonel Lee Long RFD with influencing the Army's mindset about protected vehicles thus having the unit trained in 'A' vehicle rather than bus tactics; giving greater relevance to the unit's status as a vehicle fighting force rather than a supplier of scouts.[731]

Irving also reported on further developments in the saga of the Navy Reserve review. The DRA had been critical of the Standby Reserve in the three Services and especially in the RAN. A late 2015 review of the RAN's Standby Reserve found that a number of members 'on the books' were actually deceased and contact details of about 90% of the remaining individuals were out of date. The DRA had made a detailed submission (prepared by DRA vice-president Navy, Captain Lukaitis), to the Reserve Review and made five recommendations to enhance the use of the RANR and address cultural change needed to achieve integrated equality between Permanent and Reserve members.

The DRA submission emphasised the need for essential communication tools to engage the RANR—this was picked up by the Deputy Chief of Navy when he instructed all RANR members to register on ForceNet.[732] These developments had arisen almost directly from the Lukaitis paper back in 2011. The fact that they were happening 'only' five years later was a testament to the impact of Lukaitis' original work. By DRA experience with Defence and the bureaucracy both uniformed and civil, this was a 'fast mover'. The Victorian branch of the DRA described the Review as a 'once-in-a-generation opportunity to review the Naval Reserve.'[733]

Major General Irving also noted the long-standing concern within the DRA of the rapidly declining numbers of the Reserve. The CA had tasked Major General Stephen Porter to examine recruiting and retention problems and to make recommendations to resolve these problems. As Irving quite correctly commented, however, 'this problem is now *decades old* [author's italics] and arose out of the imposition of a centralised and highly bureaucratic recruiting system on the ARes. There is no easy fix to this issue and any solution will take considerable time and resources.'[734] The DRA had fought and lost the many battles and

RAAF fire fighters Sergeant Adam Elms and Leading Aircraftman Jed Crosby dressed in their fire fighting attire with Flight Sergeant Lewis MacLennan and their Panther fire truck inside the C-130 hangar at Camp Baird in the Middle East region.
(Defence Images)

skirmishes along the way to the current system as it tried to preserve the traditional recruiting prerogatives of units and formations to recruit locally, but historically, retention was a whole other level of complexity. It would be true to say that retention issues were always present, from the very beginnings of the Association, and were directly affected by the ever-changing conditions of service for reservists experienced since 1970. The 2016 DRA conference was planned for the first time in Brisbane, at the United Services Club, on 27 August. Hoping to put a spotlight on the failure of DWP2016 to discuss Reserves roles and capabilities, the theme of the conference was set as 'DWP2016—Challenges and Opportunities for the Reserve Forces as part of the ADF'. In particular, the conference would examine just how the three Services intended to maximise the use of the Reserves. A long, drawn out election campaign and negotiation over the final government (a one seat majority for the Liberal-National Coalition under PM Malcolm B. Turnbull MHR (Liberal/Wentworth)), disrupted speaker invitations. The conference was at least able to secure a presentation from former Assistant Minister for Defence and Minister for Veterans Affairs, S. Robert and from the new Shadow Minister for Defence, Richard D. Marles MHR (Labor/Corio), who admitted to a 'steep learning curve'.

Robert highlighted both the lack of diversity in the Reserves and the ageing nature of Reserve demographics. Chief of Army, Lieutenant General Angus J. Campbell AO DSC, also spoke about ageing in the Reserve. However, as the DRA president Major General Irving pointed out in his president's message in the October 2016 edition of *The Australian Reservist*, this was in part the result of raising the compulsory retirement age of reservists several times over recent decades. [735] Raising the retirement age was perhaps a tacit acknowledgement that Reserve recruiting and retention was not sustainable, and indeed, Lieutenant General Campbell in his presentation 'admitted something that the DRA has been highlighting for years, that the Army Reserve recruitment process is broken'.[736]

Other speakers included Major General Spence, presenting both on behalf of the DRSC as well as on the current and proposed initiatives affecting the Reserves. Spence completed his posting as Head CRESD at the end of 2016 and would be replaced by Commodore Bruce Kafer. At the conference, Kafer, representing Chief of Navy, outlined the outcomes of the comprehensive review held into the Naval Reserve. Irving noted with satisfaction that 'The DRA made a significant contribution to the review, and it was pleasing to note that our recommendations about improving the people management of the Naval Reserve have been acted upon'.

Irving was not so pleased with the update on Project SUAKIN by Commodore Grant I. Ferguson, RANR, Director General Suakin, saying:

> It is extremely disappointing the earlier research that showed Reserve personnel capability would be enhanced through improved remuneration by way of reserve allowances and superannuation has been ignored in developing the various Reserve Service Categories. This has been accompanied by what the DRA considers to be quite fatuous arguments that attempt to link tax-free pay with superannuation for reservists.[737]

The speakers continued: Air Commodore Robert ('Bob') P. Rodgers, AM, CSM, representing Chief of Air Force, on the challenges facing the RAAF in transitioning to a totally integrated work force was followed by Rear Admiral Jenny Firman, Surgeon General ADF—Reserves with an overview of the important role of the ADF Health Reserves. Brigadier Michael ('Mike') Arnold CSC ADC, spoke for the Commander 2 Division on recruitment and retention improvements. Major General Melick gave an update on the major events and activities schedule for the Commemoration of the Centenary of Anzac. One of the stalwarts behind the series of successful conferences over the past decade was Neil James, president of the ADA, who attended each year to act as chair for the Open Forums. There is no doubt that James' presence was a highlight in a long day—his deep knowledge of Defence matters as well as the Reserves and the DRA itself, combined with his evident intellect and sharp focus on the critical issues, made the forums memorable and always positively received.

At the subsequent DRA AGM, Group Captain Schiller stood down as Vice-President Air Force and was replaced by Air Commodore Kathryn Dunn. WO2 Catalina Sankey was also appointed DRA treasurer, while Colonel Chris Cuneen OAM became the new DRA Queensland branch president. Branch issues were few, but Tasmania branch had suffered a 'rapid decline in numbers'. The Northern Territory branch noted that it now had few P/T reservists.[738]

Later in October 2016 DRA president Major General Irving reported on the efforts by the DRA to gain representation on the Ex-Service Organisations Round Table (ESORT). While the DRA was represented in five of the eight State or Territory based Defence Committee forums, the DRA executive considered ESORT as more effective in its advice to the Minister, if such considered information provided by the DRA on issues affecting reservists. As a result, noted Irving, and in response to a review underway by the DVA into the National Consultative Framework (NCF), the DRA made a detailed submission in support of its position that the DRA should be represented on ESORT.[739]

DVA subsequently sought further information on the 'reach' of the DRA within the Reserve community. Perhaps as a result, the DRA was asked to participate in the Female Veterans and Families Forum under DVA's sponsorship in early December 2016, to explore issues of concern to female veterans, their families, partners and widows of veterans. The DRA nominated a female Reserve veteran and the spouse of a Reserve veteran to the forum. In February 2017 immediate past president Major General Barry, followed up with the Minister for Veterans Affairs to reinforce the DRA's desire to join ESORT. It was another example of how the DRA was constantly looking for opportunities to further its effectiveness as a lobby and advocacy organisation on behalf of ADF reservists; an invitation to join ESORT was finally received in September 2017.[740]

In December 2016 the DRA had also received an invitation from the JSCFADT to make a submission to the committee's inquiry raised by the Defence Annual Report 2015-16, particularly personnel matters around Project SUAKIN and flexible workplace practices. By this time, SUAKIN had evolved into the 'ADF Total Workforce Model' (TWM). The DRA's submission reminded the Committee that extensive research by Ernst & Young under Plan SUAKIN in 2011 had shown that Reserve capability would be enhanced by through better support for employment of reservists and enhanced remuneration through allowances and superannuation. The DRA's view was that the aim of SUAKIN and the implementation of the ADF TWM 'to develop a Reserve structure, supported by 'appropriate remuneration' in turn to provide the ADF with a highly reliable core part-time workforce, has not been achieved'.[741]

In reporting to the DRA about the submission, DRA president Major General Irving stated that the DRA 'articulated that the ADF Reserves are the only part time paid workforce in the Australian community not eligible for any employer contributed superannuation and there was also a moral issue with Indigenous reservists (members of the RFSUs), not receiving any super in their P/T ADF employment. Irving also noted that there was no legislation or otherwise which prevented employees on a tax free salary to not be eligible for super and any move to tax the pay of reservists would raise 'significant equity issues'.[742] Irving summarised the DRA submission conclusion, noting that the DRA would continue to closely monitor the situation:

> ...whilst implementation of the [TWM] has raised awareness of the Reserve capability, there is a significant risk that there will be little difference in the structure and utilisation of reservists before and post-implementation of the Model. Further the ADF has not grasped the opportunity presented by SUAKIN to optimise the use of its Reserves.[743]

In early 2017, the Commander 2 Division, Major General Porter, briefed senior retired Reserve officers on the 'Army Reserve Transformation Plan' that aimed to stabilise the Army Reserve workforce into a single plan focusing on opening recruiting pathways, aligning depots and demography, and making structural arrangements to better support training of reservists.[744] No doubt there remained a certain amount of 'wait and see' among some that this was yet another plan which might yet fail to realise its ambitions. In the meantime, the DRA at least could bask in the success of its advocacy with Navy to make real changes to its management of Reserves.

April 2017 saw two important articles in *The Australian Reservist* by DRA Vice-President Navy, Captain Lukaitis. The first summarised the DRA submission to the Navy review of the RANR. He recalled that the DRA had been actively concerned with the RANR since 2010, when 312 positions from only 930 Active Reserve were abolished by Navy and DGRES–N staff were also cut (DGRES-N subsequently became DGANCR, Director General Australian Navy Cadets and Reserve). As Lukaitis expressed it:

> ...these radical cuts to the Naval Reserve affected hundreds of Naval reservists who were made redundant with little or no warning. These actions also seemed to highlight a disinterest amongst senior Naval leadership in managing Reserve careers or developing the planning of Naval Reserve capability.[745]

The DRA status paper on the subject, the 'Lukaitis paper', subsequently widely circulated, was influential in the 2015 decision by Navy to conduct a wide-ranging review into the RANR. In September 2015, the final report of that review was released in August 2016, and it made 45 recommendations, noted in the second of the articles by Lukaitis. Captain Frank Kresse

ADC RANR conducted the review. In June 2017 he wrote to Lukaitis to say, following the two articles, that: 'Importantly, the DRA submission to the Review highlighted a number of parallel issues that undermined optimisation of the Reserve workforce capability....'[746]

The DRA Conference in 2017, scheduled for 19 August in Queanbeyan, just outside Canberra, had the theme 'Reserve Forces at the Crossroads'. This later became 'Reserve Forces— Opportunity to Improve capability of the ADF' a more positive spin on the approaches to the many issues still facing the Reserve Force evolution under BEERSHEBA and the ADF Total Workforce Model. Daniel ('Dan') T. Tehan MHR (Liberal/Wannon), Minister for Defence Personnel opened the conference. He committed to ensuring the passage through Parliament of the amendments to the *Defence Reserves Service (Protection) Act*, which had been a 10-year wait. The Amendments were finally passed in November 2017. Shadow Assistant Minister for Defence Industry Mike Kelly spoke for the Opposition.

Chief of Army, Lieutenant-General Campbell, spoke to One Army, focussed on capability outcomes. Somewhat curiously, like Senator Feeney the previous year, Campbell spoke about 'moving the Reserve from a mobilisation base and being considered a separate part of the Army', as if this was still a shibboleth not yet displaced from the thinking of retired Reserve officers by the developments in policy developments of the past 20 years. Perhaps it arose in part from the examination of defence real estate assets. Campbell noted that data from the last census was being used to identify new growth corridors with a view to opening new depots; and that it required a recruiting pool of about 1000 persons aged 17-29 in order to recruit one Army Reservist.[747] On the matter of Reserve depots, he informed the conference that he would not close any depots but would focus on opening new ones. However, reservists would not parade at the 40 or so depots under threat.[748]

Other speakers included Major General Porter, Commander 2 Division; Ms. Jane F. McAloon, National Chair of the DRSC; and Major General Craig W. Orme AM CSC (later DSC), Deputy President of the Repatriation Commission within DVA. Orme noted that DVA was moving from a transactional processing model to the delivery of services to veterans from the veterans' perspective; he encouraged the DRA to develop a proposal to extend non-liability health care to reservists.[749] DVA had initiated a major transformation aimed at changing the culture of the organisation to one which is veteran-centric, updating the aged IT systems and progressively digitising DVA records, along with a new 'MyService' application process. It was also pushed by a performance audit on service delivery arising from the Senate inquiry into suicides by veterans and ex-service personnel.[750] Admiral Kafer, Head Reserve and Youth Division, gave an update on activities of his division, and of particular relevance to the DRA, noted the amendments around 'consecutive' service for determining eligibility for the DHOAS scheme appeared to be moving forward, despite the DRA contention that year of service should be cumulative.[751]

At the following AGM, DRA president, Major General Irving noted in his annual report for 2017 that:

> It has been a huge challenge organising the conference program and in encouraging reservists to attend...Literally hundreds of invitations were issued and initially there was some kick back from local units in supporting the Conference. I raised this matter with Commander 2 Div and thankfully, we subsequently received more support.[752]

His comments highlighted an enduring issue for the Association—how to attract and retain serving reservists to the Association and bring in new blood to ensure the sustainability of the leadership itself. By retirement, many senior Reserve officers had other issues to grapple with in their personal lives or other volunteer commitments and found it difficult to respond to invitations to join DRA State and National Executives. It was a problem faced by many ESOs, such as the RSL and RUSI. Even the current leadership was weary—Irving candidly noted that he was finding it 'increasingly difficult to provide the time necessary to drive the DRA and enhance our visibility'. He described the considerable workload, and the failure to date to find a successor. He called for consideration to be given to how the Executive operates to make the DRA more visible and more productive.[753]

Irving also noted that 'membership remains a matter of serious concern in all Branches'. It was almost a mantra and would be repeated again in the years ahead. This was reflected to some extent in branch reports. Victoria noted the increasing age of the membership; WA reported 169 members but only 16 serving reservists among them while the DRA treasurer reported that 'capitation [fees] appears to be reducing across the States from an apparent decline in membership being experienced in most States'.[754] Tasmania branch was in dire straits and its future was in doubt; it had been 'inactive for many years'.[755] South Australia, under Brigadier Robert ('Bob') N. Atkinson AM RFD MD, reported a more robust status with 76 members, along with a catchy analogy of the DRA being like a 'crocodile in the sun', needing to be ready to respond to whatever calls were made upon it in the ever evolving Defence environment to support Reserves.[756]

With the evolution of Defence policies affecting Reserves, the DRA had shown itself adaptable and well-informed of developments affecting reservists. The provision of services to the veteran community was no exception. The DRA had pinpointed the importance of being represented on ESORT at the national level and this was achieved in 2017, giving the DRA a voice at both national and State/Territory level, where branches were represented in five of the eight DVA consultative forums. Assisting reservists to make claims to the DVA for service-related health issues was an 'increasingly important task' for State branches. Within ESORT, the DRA was also represented on the DVA Females Veterans and Families Forum.[757]

Two Government decisions made in November 2017 affected reservists. The Government decided to provide a 'White Card' enabling mental health treatments for any member of the ADF who had served, even for one day. However, this decision had excluded reservists who had not seen full-time service, regardless of their years of service as a reservist. The DRA saw this as an equity issue and DRA president Major General Irving, as a new member of ESORT, raised it in that forum. Also, Federal and State Ministers responsible for veteran issues agreed on a common definition of 'veteran', defined as 'a person who is serving or has served in the ADF', including reservists. Following the Senate inquiry of 2017, a recommendation that the Productivity Commission to review the legislative framework and administrative processes within DVA with the object of simplifying the system weas accepted by the Government. The DRA as a member of ESORT was to participate in the review.[758]

At the ESORT meeting in March 2018, DRA president Major General Irving was able to speak directly to Darren Chester (the third Minister for Veterans Affairs and Minister for Defence Personnel in as many months). Among the matters discussed Irving noted to the Minister that recent Defence press releases from the Minister's office noted that in giving numbers of serving personnel and veterans, the Minister's office had completely ignored 25,000 reservists and thousands of ADF cadets, both of which fell under the Minister's responsibility.

The DRA conference for 2018 was held on 25 August in Canberra, with the theme 'Meeting the Challenges for Reservists to Serve'. The conference examined not just the critical issues associated with meeting those challenges but also the actions required of Government, Department of Defence, and employers to enhance the availability of reservists. Unfortunately, due to the political upheavals within the Government, notably leadership spills in the Liberal Party, which saw Malcolm Turnbull ousted as PM, being replaced by Scott J. Morrison MHR (Liberal/Cook), some Government speakers could not attend. However, Gai M. Brodtmann MHR (Labor/Canberra), the Shadow Assistant Minister for Cyber Security and Defence, did attend. She was described as 'a strong supporter of the DRA'. Major General Porter once again attended as Commander 2 Division to report on progress with the Army Reserve Transformation, especially around the development of training.[759]

Jane McAloon and Rear Admiral Kafer also joined the conference again. Both Porter and Kafer raised the question whether the term 'reservist' was divisive and perhaps should even be 'deleted from our lexicon'.⁷⁶⁰ Speakers included Brigadier Simon C. Gould DSC, SUAKIN Branch/People Capability Division, who provided an update on the ADF's Total Workforce Management strategy (the former Project SUAKIN). Group Captain Tony Hindmarsh CSC, Project Officer/ DGR-AF, briefed the conference attendees on the Department of Defence Joint Reserve Working Group review of the current employment package for Active Reserves and also an update on best utilization of Reserves in the Air Force.

Brigadier Michael ('Mike') H. Annett CSC, recently returned from active service in Afghanistan and as DGR-Army– and a future DRA president—gave an optimistic report regarding improved numbers for the Reserve due to lower separation rates and improved recruitment figures, which essentially indicated a stabilisation of the Reserve workforce. Commodore Mark D. Hill, CSC RANR, DGANCR, provided an update on progress in implementing Navy Reserve recommendations arising from the review of 2016.⁷⁶¹

Of particular interest was a presentation by the DRA Vice President—Army, Major General Wilson, reviewing the DRA's impact on policy development over the past ten years. He referred to the 10 ideas for an independent review of the Reserve, articulated by Major General Garde when Garde was DRA Victoria president at the 2010 DRA conference. A number of these ideas had been supported by the DRA to fruition under the current Plan BEERSHEBA for example, a plan which has 'given the Army reserve a new sense of purpose'. Wilson noted that the DRA had made major submissions on Reserve matters to each of the Defence White Papers since 2009. This effort reminded the politicians and service chiefs of the Reserve viewpoint. For some years, the service chiefs have acknowledged that they cannot operate effectively without significant input from their respective Reserves. Enhanced capability for Reserves has also been a catch cry of the DRA for many years, and it has influenced the thinking of Defence in this regard, both with Plans BEERSHEBA and SUAKIN.⁷⁶²

Wilson went on to list the issues which remained on the DRA's 'radar', both nationally and regionally. He noted that although the amendments to the Defence Reserve Service (Protection) Act had finally been passed in November 2017, and that this took an incredible 10 years to achieve, the DRA had nonetheless not given up on these important changes and had continued to advocate year in and year out. The DRA, said Wilson, 'is patient if nothing else'! Superannuation and tax-free status of Reserve pay remained on the table. Wilson noted that when tax-free pay was last taxed in 1983, Reserve numbers dropped from 33,131 to 23,722 virtually overnight, and Reserve numbers never recovered.

Although the decision was later reversed, Reserves remains the only group of employees that are not entitled to superannuation; an issue very much on the DRA's list of issues.

Leave entitlements and other conditions of service are related to the mode of employment—reservists remained as a 'casual' as distinct from 'part-time' work force—but the 'daily rate of pay' system which Defence uses to calculate entitlements reinforces the status. The DRA also continues to lobby for hypothecation of Reserve expenditure. along with a more expeditious recruiting process and recognition of reservists in Awards.[763]

Wilson ended his presentation with a proposal to challenge the gap between employing reservists under part-time Conditions of Service as opposed the current casual Conditions of Service:

> ...the current remuneration system for the part-time components of all three services is outmoded and completely inappropriate. Rather than 'fiddle around the fringes' with such issues as superannuation entitlements, eligibility for the Defence Home Owners Assistance Scheme, the 365 day divisor and leave entitlements, it is proposed that a 'grass roots' review of the remuneration system for the part-time component be undertaken by a working group comprised of representatives of the DRA, Reserve and Youth Division, the Directors General—Reserves of each service and the Defence People Group.[764]

The DRA AGM was held the following day. The Queensland branch reported 50 members.[765] Like many of the DRA branches, it was busy with events and activities which helped to raise the DRA profile in the Defence community, but in general there were few activities of the branch committee which contributed to the DRA's policy positions overall. This was not a criticism of the committee; on the contrary, it was well represented at committee level, and it was working hard. It was more of a reflection of just how successful the DRA had been in addressing many of the major issues which had been to the fore in the past decade. A reflection, in fact, of how many had been resolved as Defence policy towards Reserves had evolved over the past two decades along with a more evident understanding and consideration of Reserves in the context of ADF capability needs since Operation CITADEL in East Timor.

By April 2019, the DRA president Major General Irving was able to report that for several months the DRA had been busy with a large number of inquiries and proposed legislative changes affecting veterans—which now included reservists. The immediate past president, Major General Barry, also represented the DRA on the DFWA and Irving on the national council of ESORT as well as the Alliance of Defence Service Organisations (ADSO). Along with branch representation on a number of DVA Consultative Forums, the DRA was providing through these roles submissions to Government, the Productivity Commission, National Audit Office, People Group in Defence and the DVA itself on a range of issues affecting reservists.

For example, the DRA advocated that all reservists who had one day's service in the ADF should be entitled to mental health treatment without the need to link the condition to the reservist's ADF service (non-liability health care). The 2018 Government budget had extended mental health care to those reservists who had rendered Reserve Service Days with disaster relief, border protection or involvement in a serious service-related accident. The DRA hoped this would be extended to all reservists in the 2019 budget.[766]

Other benefits affecting reservists (but not all) had come through. In October 2018, the Government launched an Australian Defence Veterans Covenant to provide recognition of military and in February 2019 the Military Rehabilitation and Compensation Amendment (Single Treatment Pathway) Bill was introduced to parliament. This bill replaced the dual treatment pathway model under the MRCA 2004 with the new single pathway. MRCA 'clients' would then be able to access the new Veteran's Card and no longer have to pay up front for treatment.

The recommendations of the Senate Foreign Affairs, Defence and Trade References Committee into veteran suicides were still being implemented in 2019 and the Productivity Commission subsequently reviewed veteran compensation and rehabilitation. ADSO, including DRA representation, made a submission to that inquiry and a final report was due in July 2019. Some seemed surprised to learn that Reserve enlistment and discharge information had never been provided to DVA; this was to change.

For almost the full 50 years of the Association's existence in its many forms, the apparent discrimination against reservists being honoured in Queen's Birthday and Australia Day awards lists had long been a subject for complaint by the Reserves. There were many ideas about why this was, including the failure of commanding officers to submit quality recommendations, or any at all. Even in recent years, awards to reservists had been sparse. In 2017 for example, there was not one award. In 2018 there was one but unusually in January 2019 there were no less than 21, although only five were military awards. Nonetheless, the others also recognised the civilian contributions to the community and Australian society by those who were also members of the Reserve, thus demonstrating the true dual nature of Reserve service.

A federal election in May 2019 led to new appointments in the third-time elected Liberal/National coalition. Senator Linda K. Reynolds CSC (Liberal/Western Australia) was appointed Minister for Defence. A former brigadier in the Army Reserve, Reynolds was the first woman to reach that rank. Darren Chester was appointed Minister for Veterans Affairs and for Defence Personnel while Alexander ('Alex') G. Hawke MHR (Liberal/Mitchell) was made Assistant Defence Minister. Melissa L. Price MHR (Liberal/Durack) was appointed

Minister for Defence Industry. On the opposition benches, Richard Marles became Shadow Minister for Veterans' Affairs and Defence Personnel and Shayne K. Neumann MHR (Labor/Blair) was appointed as Shadow Minister for Defence.

For the 2019 DRA Conference held on 17 August in Brisbane. The theme was 'Reserve Forces: Building on the Success of the Total Force'. Reynolds and Neumann represented the Government and Opposition respectively. While quite new in their roles, Reynolds in particular was also able to draw upon her Reserve experience to demonstrate why she was a 'quick study' for her presentation. The Governor of Queensland, Paul de Jersey AC QC, opened the conference—like Reynolds, he was also a former Reservist.[767] Rear Admiral Brett S. Wolski AM, who had left full time service in February 2019 and moved to the Reserve with the appointment as Head Reserve and Youth Division and Commander ADF Cadets, gave an update on his division and its activities in the context of TWM (the former Plan SUAKIN) in particular. He was joined by Commodore David Greaves, DGANCR.[768] The TWM in Army was addressed by Major General Kathryn J. Campbell AO CSC, Commander 2 Division, while Group Captain Joanna ('Jo') Elkington provided an update on Air Force Reserves on behalf of the DGR-AF.[769]

Jane McAloon, Chair of the Defence Reserves Support Council returned to the conference for the third time. She explained that an independent review of DRSC and Defence's Reserve and Youth Division by consultants KPMG had been commissioned in March 2019 to 'understand the implications for civilian employer engagement of the changing strategic requirements for defence'. This followed the revelation in earlier surveys that DRSC did not know either which employers or industries employed actual reservists!

The review, along with a SERCAT-3-5 member survey revealed that almost 40% of reservists were employed by emergency services across Australia, and that the next largest category were actually self-employed reservists.[770] This was the second KPMG review; the first had been undertaken in 2017 but the DRA had not been consulted; neither was it consulted for the 2019 review. DRA president Major General Irving (and the DRA's representative on the DRSC nationally) expressed his 'deep concern at [this] failure and was subsequently able to 'articulate the DRA view on the role and function of the DRSC'.[771]

> Irving noted:
>
> > Despite the promise of widespread consultation, the DRA was not consulted on this review. I then wrote to the Chair of DRSC, Jane McAloon, expressing our deep concern at the failure, once again, to consult with the DRA. This resulted in apologies and contact with KPMG where I was able to articulate the DRA view on the role and function of the DRSC.

He went on to say: 'It is vital with any proposed restructure of the DRSC that the DRA remains on the National DRSC'.[772] It did.

At the AGM following the conference, a number of changes were made. For the first time in many years, the Association appointed a Patron, Jack Smorgon AO, former Chair of the DRSC. Brigadier Phillip K. H. Bridie AM, the Director of RSL NSW and Honorary Colonel of 1/15 Royal NSW Lancers was elected to replace Major General Wilson as Vice-President—Army. Lieutenant Colonel Laureen Grimes, who was the first female chair of the Victorian Veterans Council, was appointed as National Secretary, replacing Lieutenant Colonel Cotton. In the summary of the 20 action items arising from the AGM, no less than 16 concerned issues internal to the organisation of the DRA itself. Of the remaining four, only two were 'issues' and they were not new.[773]

It was perhaps a reflection of the challenge facing the DRA of remaining relevant and engaged at the national level on issues affecting the Reserves. Certainly, 'The relevance of the DRA from the perspective of serving reservists appears to have been diminished by the widespread acceptance of the Total Force concept and the deployment of reserve forces overseas and within Australia'.[774] The DRA Victoria report for 2018/19 noted the dilemma. On the one hand—'...the development of the Total Force is proving ... successful and also has significantly met the expectations of serving reservists to contribute to defence in meaningful roles'. On the other: 'This has however placed substantial demands on the amount of time members are expected to serve and is in turn putting pressure on developing and retaining experienced members'.[775] Many volunteer organisations had faced similar dilemmas with their own constituencies, when major issues appeared to have been resolved or objectives met, but once again the working environment for reservists was about to change.

Only a month after the AGM, the bushfire 'season' began with small fires at first but developed in late December into widespread and devasting fires across several States. In September 2019 the CDF had warned in a speech to managers of Government departments and agencies that the ADF would be stretched in the future by increasingly frequent natural disasters caused by climate change along with more conflicts around the world.[776] Events to come bore out this prediction.

The expanded role of the Army Reserve in domestic operations was identified and enhanced by the decision by the Chief of Army on 1 October 2020 to formally assign domestic operations, less counter terrorism, to the 2nd Division. Operation BUSHFIRE ASSIST, the callout of reservists and permanent ADF to support bushfire operations by local and State emergency services across several States between December 2020 through March 2021, once again highlighted the versatility and utility of the Reserves.

The ADF at its peak was providing 6500 members to the Operation, including 3000 reservists. The legal limitations on calling out the ADF in support of national domestic emergencies however became apparent and was further highlighted when once again 2200 ADF members including reservists were called out to help State authorities manage quarantine and border issues arising from the COVID-19 pandemic which began to affect Australia as early as January 2021.

For the Reserves, and the ADF in general, the legislative provisions of its call-out for bushfire and COVID-19 operations—in the latter case which continued throughout the year—was recognised as somewhat inflexible. The Government subsequently announced plans to amend the Defence Act to allow the Prime Minister to declare a national emergency or disaster and deploy the ADF within Australia. The proposed *Legislation Amendment (Enhancement of Defence Force Response to Emergencies) Bill 2020* was introduced to Parliament in early September 2020, and aimed to:

- streamline the calling out of the Reserves.
- decouples reserve call out from the requirement for Continuous Full-Time Service (CFTS) by allowing the CDF to direct a Reservist's service on the basis of Reserve Service Days (RSD) or CFTS.

RAAF Leading Air Craftsmen Clay Brown (left) and Matthew Kirk from No. 65 Squadron assess the ground whilst cutting a fire break in the Bando Forest to protect pine plantations near Tumut, NSW in January 2020 during the national bushfire emergency over 2019–20.
(Defence Images)

- provides for superannuation to reservists called out on CFTS—backdated to November 2019; and
- provides a level of personal immunity to reservists from criminal or civil liability whilst on call out, like that provided to rural bushfire and other civilian emergency workers.

DRA President Major General Irving later reported:

> Sadly, no one from Defence or the offices of the [Minister for Defence or Ministry for Defence Personnel] contacted the DRA to advise that the Bill had been introduced into Parliament. In fact, the DRA was advised of the Bill by the office of the [Shadow Minister for Defence Personnel]. As a consequence, I wrote to the [Minister for Defence] on 25 September about the failure to consult with the DRA in relation to the Bill.
>
> On 10 October, the [Minister for Defence] replied, stating *inter alia*, "My sincere apologies that your organization was not consulted in the drafting of this Bill. I have asked Defence to remedy this…" As a result of my letter, senior members of Defence contacted me and a level of consultation occurred. However, the point remains, it was three weeks after the Bill was tabled in Parliament before Defence contacted me and then only because I had complained to the [Minister for Defence] about lack of consultation.[777]

In early October 2020, the Senate referred the provisions of the *Defence Legislation Amendment (Enhancement of Defence Force Response to Emergencies) Bill 2020* to the Senate Foreign Affairs, Defence and Trade Legislation Committee for inquiry and report by 4 November 2020. Some 58 submissions were made to the Inquiry. The DRA made a submission on 15 October 2020 and DRA President Major General Irving gave evidence before the Committee on 30 October. The DRA submission to the inquiry supported the proposed amendments overall as 'a positive step for reservists and Reserve service.'[778]

The backdating of superannuation benefits to called out reservists to November 2019 and the provision of a level of personal immunity to reservists whilst on call out duty were also fully supported. Major General Irving's submission for the DRA noted that the amendments 'allows greater flexibility in call out procedures, is administratively easier and is probably less costly for Defence'. However, the DRA submission also noted the importance of actual records of service of reservists, not least due to exposure in terms of protection and cover if killed or injured whilst on that service. The DRA's call for better and detailed records of Service was directed mainly at Defence, rather than the legislation, but the linkage of this important requirement 'needed to be addressed'.[779] The Senate inquiry, by a majority, recommended that the Bill be passed without delay, subject to the Explanatory Memorandum being amended to clarify the intention and operation of the Bill.

Meanwhile the COVID-19 pandemic affected the operations of the DRA both nationally and regionally throughout 2020. Face-to-face committee and executive meetings could not be conducted and reverted to technology in the form of video 'zoom' discussions. Travel was curtailed and briefings and liaison with the various ESOs and other connections had to adapt to similar measures. There was no edition of *The Australian Reservist* in April, although that was impacted in part by the publisher being unable to develop sufficient advertising support at a time when there was a broad business contraction across the country due to the pandemic.

One major impact of the pandemic, however, was on the planning, timing, and conduct of the national conference. It was just not possible to hold the conference in Adelaide in August as originally planned. Due to the ever-changeable situation regarding the opening of State borders, the DRA Executive decided to conduct its 2020 national conference by way of a video format. Thirteen speakers provided videos and members were given an opportunity to ask questions on-line. Despite concerns about this new way of conducting the conference, statistics post conference pointed to its overwhelming success, that the day after the conference link went live, there were over 4000 hits on the link, including 145 from overseas countries. The speaker list, was, as always, impressive:

Army Reserve soldier, Private Jac Harman from 13 Combat Service Support Battalion, assists a Western Australia Police officer, to conduct a traffic stop at a police control point at Green Head, WA, under Operation COVID-19 ASSIST.
(Defence Images)

- Darren Chester, Minister for Defence Personnel, representing the Minister for Defence
- Shayne Neumann, Shadow Minister for Defence Personnel, representing the Shadow Minister for Defence
- Major General Roger J. Noble DSC AM CSC, Head, Military Strategic Commitments Division
- Major General Natasha A. Fox AM CSC, Head People Capability
- Elizabeth ('Liz') Cosson AM CSC, Secretary of the Department of Veterans' Affairs (former Major General)
- Major General K. Campbell AO CSC, Commander 2nd Division
- Rear Admiral B. Wolski, AM RAN, Head, Joint Support Services Division
- Brigadier Duncan Hayward CSC, Director General Defence Force Recruiting
- Commodore Michael Slattery RAN, Director General Australian Navy Cadets and Reserves
- Brigadier Douglas ('Doug') W. Laidlaw CSC, Director General Reserves—Army
- Air Commodore Tony Hindmarsh CSC, Director General Workforce Design and Reserves—Air Force, and
- Lieutenant Colonel Andrew White, Commanding Officer 1/15th Royal New South Wales Lancers

At the 2020 AGM, DRA president Major General Irving, presented his report for the year. While noting that the year had been 'a very difficult year for our members' due to the effects of the pandemic on a range of activities such as Anzac Day, Irving also pronounced the year as 'an unprecedented year for the ADF Reserves', citing reservist support for Operations BUSHFIRE ASSIST and COVID-19 ASSIST. Irving also noted:

> There is a risk, with the potential for more frequent call out of reservists on disaster relief and associated operations, that their families and employment could suffer. There is also a potential loss of war fighting skills. The DRA will need to closely monitor this into the future.[780]

After reporting on the *Legislation Amendment (Enhancement of Defence Force Response to Emergencies) Bill 2020*, Irving turned to another issue that the DRA was looking at. The DRA, together with other veterans' organisations, had for many years advocated for a veterans' covenant to recognise the unique nature of military service. In October 2019 legislation passed Federal Parliament, introducing the Australian Defence Veterans' Covenant.

The Government, through DVA, decided to issue a wider recognition package including a lapel pin, oath and supporting veteran card to any member or former member of the ADF

Australian Army Reservist, Lieutenant Bartholomew Arundell (L) in the Intensive Care Unit at the Box Hill Hospital, Victoria during Operation COVID-19 ASSIST.
(Defence Images)

who had completed one day's continuous full-time service. A different lapel pin, but no veteran card, is being provided by DVA to ADF reservists who have had no period of full-time service, with some limited exceptions. Reservists ineligible for a veteran card, including many long-serving reservists, are denied the opportunity to access the business benefits associated with the veteran card. As Irving pointed out in his report: 'Issuing a different lapel pin to Defence reservists, and without a veteran card, is both divisive and contrary to the aims of a totally integrated ADF'.[781]

The DRA president concluded his report with some minor order, but important, organisational considerations. His workload as president, while still considerable has been aided in the past year by the appointment of a national secretary. He noted in a now familiar but no less perturbing refrain that 'Membership still remains a matter of serious concern in all Branches. There is no easy answer to increasing our membership. It will take much creative and hard work by all Branches and the National Executive'. And he added: 'We need to give careful consideration to both the future of and format of *The Australian Reservist*—hard copy or electronic, or combination'. In part this was a result of technology changes as much as concerns about advertising support and the demographics of the readership. Indeed, Irving reported that he would continue to meet with the DRA National Executive and State Branch Presidents using online conference technology—'This initiative has greatly enhanced communications on a wide range of issues affecting reservists'.[782]

As 2020 ended and along with it, the first fifty years of the Association, it was technology which had transformed the ADF and the Reserves more than any other single factor due to the accelerated impact was having on the way armed conflict was conducted and increasingly determined by, technology capabilities. The DRA in its own way was also affected by technology as was its membership. As one observer noted, in the past thirty years especially, 'our lifestyles, workplaces and economy have been dramatically transformed'. For example in 1990, only one percent of us had a mobile phone, and they were ...not even smart enough to receive text messages.'[783] Hence the challenge for the DRA into the future, keeping up with change affecting the Reserves and utilizing technology to remain in touch and relevant to its members. 2020, for all of its challenges, seems to have shown a possible pathway ahead.

Epilogue
The Sum of the Parts

Following the ups and downs and ebbs and flows of any national volunteer organisation over a fifty-year period is challenging at the best of times. Following the DRA and its predecessors the CMF and then Army Reserve Associations—through what can only be described as tumultuous change affecting the Reserves over this time—is even more so. On average the Reserves have been buffeted by the outcomes of reviews, reorganisations, and restructures in every single year of those 50 years from 1970-2020. The resilience of both the reservists themselves and the representative organisation, the DRA, in face of this change and sometimes existential threats to the very existence of Reserves, has been remarkable.

At the beginning, with the first formal meeting of a small committee in March 1970, the Association as such was the work of mainly one person, Major General Paul Cullen. He and a small coterie of supporters, many of them fellow veterans of World War II, persevered against opposition from Regular officers and Reserve officers alike to get an Association on its feet. The CMF Association was initially NSW-centric, a reflection of the sheer *gravitas* that NSW had within the Reserve organisation across the country. Its 2nd Division outlived Victoria's 3rd Division and remains the pre-eminent Reserve formation. However, as the Association expanded gradually to States other than NSW it became clear that there was considerable interest in and enthusiasm for, an organisation that could speak up for reservists and to try and preserve its traditions and capabilities, not least against Regular Army discrimination and antagonism and political neglect.

The Association continued to evolve and respond to changes, changes often to the detriment of Reserves. A formal history of the Reserves themselves post 1974 remains to be written, although this history of the Association reflects and touches on many of the keynote changes affecting the Reserves over the period to 2020. It is therefore easy to forget, within the context of these changes (some of them fundamentally affecting the Reserves) that the Association never gave up trying to influence the processes and pace of change for the betterment of the Reserves.

In the earlier years of its existence, the Association's senior members such as Cullen, managed to influence directions and outcomes for Reserves through direct representation to senior politicians of the major parties. Despite mutual suspicions and a certain level of cynicism at times, Association members also gradually came to an understanding with Regular officers that the Reserves had an important role to play even when cuts to resources and finances put the Regular Army at odds with the Reserves. It took many years for this discord to be overcome.

In 1970, the CMF Association arose from the direct experience of the founders of the Association from World War II and then the concerns of the Cold war which followed. It was a central part of thinking of the Association that the Reserves were a mobilisation force needed to be on standby in case of a national security crisis; an expansion force critical to the capability of Australia to defend itself. The impact of the Vietnam War and the apparent lack of direct threats led this thinking to be largely discarded over time. The Reserves were perceived to be increasingly irrelevant given short lead times, having a low level of capability in the context of the growing sophistication of the battlegrounds of the future. One wonders whether, if no Association existed to support them and advocate for them, whether the Reserves might have disappeared altogether. It is clear from the internal accounts of the Association that at times its leadership despaired that anyone was listening to them.

The Association itself also had to change along with some of the views of its leadership team. Some State branch rivalries exposed the need to become more professional and national in its approach in order to develop detailed and lengthy submissions at both branch and National Executive level to the plethora of Defence reviews, reports, and Parliamentary inquiries through the 1980s and into the 1990s. It was noticeable that this also put considerable strains on the Association's resources and capabilities to respond. Members had to consider not just increasingly complex issues affecting Reserves but also Tri-Service Reserve issues at that. The Association evolved, somewhat belatedly, to encompass all three services and organisational and structural changes and became the DRA in 1992. Yet at times it seemed tired, and some individuals were burned out by the burden of hard work with little apparent reward or recognition for reservists.

However, throughout these first 25 years the Association persevered to make itself heard and remain relevant on issues affecting the Reserves. No other organisation could do so. The impact on specific issues was hard to measure. Yet there is no doubt that the weight of its persistent advocacy, the determination of the Association leaders both national and at branch level, and the internal strength it derived from the seemingly endless committee meetings, briefings, speeches, and gatherings of interested members at annual general and branch meetings over these long years created a momentum.

The Association was not flamboyant or loud. It just continued to work towards its objectives to the point that it could not be ignored or sidelined. While some personalities and at times some Association branches were more prominent than others, what shines through is the Association team work, the energizing effect of having large problems to solve and the motivation to offer solutions to enhance and improve the Reserves. It should not be surprising. After all, many of the retired officers and senior NCOs were also highly capable individuals in their civilian lives as well.

Subsequently, there was enormous progress for Reserves especially between 1995 and 2005, with the catalyst being the East Timor crisis in 1999. It was a shock to the Army in particular, as it struggled to put even a battalion group into an operational theatre, let alone a brigade. Suddenly there was a new focus on the utility of Reserves. Over time this focus sharpened rather than decreased, as operational deployment of the ADF to a variety of operational settings began to ramp up. The creation of a High Readiness Reserve within the Active Reserve highlighted deficiencies in Reserve recruiting, training, capability, and capacity to support permanent ADF units in the field. Gradually this resulted in a rapid evolution in the thinking—from a Networked and Hardened Army to One Army to the Total Force Concept. Force Modernisation in turn led to both Plans BEERSHEBA and SUAKIN, in which the DRA had become an important constituency to consult.

The DRA continued to adapt to these evolutions in the strategic environment as well as in the ADF itself while turning its attention to 'lower order but still important issues', especially around conditions of service.[784] It struggled at times to be relevant to Air Force and Naval Reserves, especially as these Services had different models for Reserve service than Army, although it proved itself to be a major influence behind the commissioning of a comprehensive review of Naval Reserves completed in 2016.

Its efforts to put in place vice-presidents for each of the Services plus health services was an effective response to recognising the DRA needed to address all Service Reserve issues, not just those affecting Army. The Reserves were also changing—fewer Reserve positions at senior level, integration of reservists and Reserve units into ADF formations, more deployments for reservists overseas on continuous full-time service, more ex-permanent ADF members becoming reservists, changes to training requirements and conditions of service and with the Total Force Concept, greater flexibility in service all round.

For the DRA itself, an ageing demographic in its membership, fewer senior retired officers willing to join in the DRA, a dearth of serving reservists as members and the decline of formerly thriving State branches were countered to some degree by the growing importance of its national conferences as the main activity of the Association each year. The conferences, improvements to the Association journal, *The Australian Reservist* and a website were all important to the DRA continuing to get its messages out.

The DRA conferences also became the only series of annual conferences in Australia to focus exclusively on Reserve matters—and the Reserves mattered more to ADF capability by 2020 more than at any other time in its history, perhaps since World war II. Increasingly, politicians and senior Defence officers from the CDF down also saw the DRA conferences as vital to put forward their views of developments and to elicit Reserve support for those developments.

The theme of the 2020 DRA conference—'*Quo Vadis*: An Integrated Future'—seems very apt after 50 years. Several times during these long years of its existence the Association has asked this extremely healthy question of itself. The Association could declare 'victory' and accept that the word 'Reserve' in the evolved and integrated Total Force, which is the ADF today, has become an anachronism. The Navy, after all, has abolished the term from its lexicon already.

Yet at the very time when the DRA appears to have reached ascendancy in its advocacy from 50 years of work, with the Reserves in a strong position of capability contribution and many if not most of the issues around Reserve conditions of service settled, some big questions have returned. The challenges to the ADF outlined by the CDF in his 2019 speech regarding the increasing array of potential tasks facing the ADF in terms of both security and climate change emergencies were underscored in dramatic fashion by the PM, Scott Morrison in mid-2020.

Announcing a $270 billion spend on defence through 2030, Morrison warned that 'Australia faces regional challenges on a scale not seen since World War II'. The defence strategy focuses on the Indo-Pacific region and Australia, says Morrison, has to prepare for a post-COVID-19 world which is 'poorer, more dangerous and more disorderly.'[785] Somewhat presciently, the author of *The Once and Future Army* wrote in his conclusion to that history of the CMF to 1974: 'History may yet provide the opportunity for *Citizen Soldiers* to again show their worth on a large scale'.[786]

While the 2020 Government announcements did not presage a large increase in the manpower of the three Services, given that the emphasis was on further enhancements of the ADF's reach and capabilities, there can be little doubt that the ADF will have to expand its numbers to meet these external and domestic challenges both physically and technologically. The Reserve will continue to be a vital part of the new paradigms and the DRA will continue to be at its side for years ahead. As an organisation it will have to continue to adapt if it is to remain contemporary and relevant to today's reservists. Its ongoing efforts to rejuvenate and energise its membership base are as important as ever as are succession plans to maintain focus at both Branch and National Executive level.

One wonders what Major General Cullen would make of the Association now. He would instantly recognise the new regional security issues as relevant to the Reserves and the reservists themselves as an important part of Australia's defence preparedness and capability. There might be fewer than in his day, but they are unquestionably far more capable in their contribution, while being no less patriotic. He might deplore the loss of traditional Reserve units and traditions which has occurred over the years, but equally applaud the apparent end of the old divisions between Regular (permanent) and Reserve as well as the growth in Reserve capability and deployments within the Total Force concept.

Cullen would, no doubt, be pleased to see the number of reservists with operational experience, and deployment of Reserve units. He would see an Association still able to advocate for Reserves in a way which wasn't possible in 1970, an Association still closely engaged with employers and liaising and coordinating closely with the now formidable range of ESOs to represent Reserve interests to Government and Defence interests alike. Major General Cullen started something in 1970 which has gone from strength to strength, despite the setbacks, disappointments, and the opposition to it which the Association has endured. The status of the Association today, in all its forms over the past 50 years—from CMF Association to Army Reserves Association to Defence Reserves Association—is a testament to its leadership, its members and the Reserves themselves.

Annex A
CMF Association 1970-1986

National office bearers

Note: In some States, there was no formal president for years after 1970, but there were nominated representatives

President
- Major General P.A. Cullen CBE DSO ED, 1970
- Major General A.C. Murchison AO MC ED, 1975
- Major General J.M.L. Macdonald AO MBE ED, 1979
- Major General G.L. Maitland AO OBE RFD ED, 1983
- Brigadier I.H. Lowen OBE ED, 1984

Deputy President
- Lieutenant Colonel B. Nebenzahl, 1985
- Colonel L.A. Simpson CBE, 1986

Executive Director
- Major G.E. Cory MC DCM, 1970-1973

Secretary
- Lieutenant Colonel J.R. Dart OBE RFD ED, 1970
- Colonel W.E. Glenny ED, 1978
- Lieutenant Colonel D.I. MacLeod RFD ED, 1985

Treasurer
- Colonel D. Pennicook ED, 1970
- Lieutenant Colonel M.W. Fairless ED, 1973
- Lieutenant Colonel R.M. Faulks ED, 1979
- Brigadier J.R. Dart, 1982
- Colonel D. Sandow RFD ED, 1985

State presidents

New South Wales
    Colonel T.J. Crawford ISO MVO MBE ED, 1976
    Major General A.C. Murchison, 1984
    Brigadier R.S.P. Amos RFD ED, 1985

Northern Territory
    Captain J. Dunn, 1976
    Lieutenant Colonel D.G.B. Prosser MBE ED, 1979
    Major J. Dunn RFD, 1986

Queensland
    Brigadier M.E. Just OBE ED, 1976
    Brigadier J. Springhall MBE ED, 1979
    Brigadier K.B. Whiting AM ED, 1981
    Brigadier R.I. Harrison MBE RFD ED, 1986
    Committee formed 1986

South Australia
    Brigadier Sir Thomas Eastick CMG DSO ED, 1976
    Brigadier J.G. McKinna CMG CBE DSO LVO ED, 1981
    Colonel R.J. Stanley CBE ED, 1986
    Committee formed 1986

Tasmania
    Colonel L.A. Simpson CBE RFD ED, 1976
    Lieutenant Colonel A.C. Bidgood RFD ED, 1986

Victoria
    Committee formed 1977
    Brigadier I.H. Lowen, 1979
    Lieutenant Colonel D.E.F. Bullard OBE RFD ED, 1986

Western Australia
    Brigadier J.B. Roberts MBE ED, 1976
    Colonel R.D. Mercer AM RFD ED, 1981
    Lieutenant Colonel T.J. Arbuckle RFD ED JP, 1986

Editor: *Citizen Soldier*

   Lieutenant Colonel A.C. Browne ED, 1972

   Lieutenant Colonel B. Nebenzahl RFD, 1986

Publisher: Citizen Soldier

   Major B. Nebenzahl, 1972

Annex B
Army Reserve Association
1987-1992

National Office Bearers

President

Brigadier I.H. Lowen on leave
Brigadier R.I. Harrison, 1987
Brigadier R.S.P. Amos, 1990
Major General R.G. Fay AO RFD ED, 1992

Deputy President

Brigadier R.I. Harrison, 1987
Major General K.G. Cooke AO RFD ED, 1990

Secretary

Lieutenant Colonel D.H. MacLeod, 1987
Colonel K. O'Dempsey RFD, 1988
Major P.R. Donovan AM RFD ED, 1991

Treasurer

Colonel D.J. Sandow, 1987
Lieutenant Colonel R.J. Byrne RFD, 1988
Lieutenant Colonel R. Hill, 1990
Lieutenant Colonel R.S.J. Thirwell RFD ED, 1991

Editor: *Citizen Soldier*

Lieutenant Colonel B. Nebenzahl, 1988-

State presidents

New South Wales

Brigadier R.S.P. Amos, 1987
Colonel J.S. Haynes OAM, 1988
Major General R.G. Fay, 1990
Major General J.D. Keldie AO MC, 1992

Northern Territory
   Lieutenant Colonel J. Dunn AM RFD, 1991

Queensland
   Lieutenant Colonel N. Williams, 1989?
   Colonel K. O'Dempsey RFD, 1991

South Australia
   Brigadier I.F. Barr, RFD ED, 1987
   Lieutenant Colonel M.E. Alexander RFD ED, 1990

Tasmania
   Lieutenant Colonel A.C. Bidgood, 1991

Victoria
   Colonel B. D. Clendinnen AM RFD ED, 1987
   Lt. Colonel D.E.F. Bullard, 1990
   Colonel B D Clendinnen, 1991

Western Australia
   Lieutenant Colonel E Quartermaine

ANNEX C
Defence Reserves Association 1993-2020

National Office Bearers

President

    Major General R.G. Fay AO RFD ED, 1993

    Colonel Donald J. Sandow RFD ED, 1994-1997

    Major General W.E. Glenny AO RFD ED, 1997-2005

    Major General J.E. Barry AM MBE RFD ED, 2005-2013

    Major General P. Irving AM PSM RFD, 2013-2020

Deputy President

    Major General K.G. Cooke AO RFD ED, 1993

    Major General K.G. Cooke, 1994-1995

    Major General W.E. Glenny, 1995-1997

    Brigadier A.J. McGilliard RFD ED, 1997-2007

    Major General P. Irving, 2010-2013

    Major General I.B. Flawith AO CSC, 2014-2020

Secretary

    Major P.R. Donovan AM RFD ED, 1993

    Lieutenant Colonel I.D. George RFD ED, 1994-1996

    Flight Lieutenant H. Quelch, 1997-2000

    Lieutenant Colonel G.J. Sharkey RFD, 2001-2007

    Captain P.W. Wertheimer OAM RFD, 2008-2010

    Major E.D. Bedggood, 2011-2018

    Lieutenant Colonel L. Grimes, 2019-2020

Treasurer

    Lieutenant Colonel R.S.J. Thirwell RFD ED, 1991-1993

    Lieutenant Colonel J.F. Henry RFD ED, 1994-1996

Lieutenant Colonel P. Hall, 1997
Lieutenant Colonel J. Moore, RFD, ED and
Lieutenant Colonel P. Edwards RFD, 1998-1999
Lieutenant Colonel P. Hall RFD, 2000
Major R. Millbank RFD JP, 2001-2005
Colonel D. Townsend RFD, 2006-2010
Lieutenant Colonel J. Cotton RFD, 2011-2018
WO2 C. Sankey, 2019-2020

Vice-President -Navy

Commodore R. Elsey RFD RANR, 2011-2013
Captain J.L. Lukaitis AM RFD RANR, 2014-2020

Vice-President—Army

Major General I.B. Flawith AO CSC, 2011-2013
Major General N. Wilson AM RFD, 2014-2018
Brigadier P.K.H. Bridie AM, 2019-2020

Vice-President—Air Force

Air Commodore P. McDermott AM CSC, 2011-2013
Group Captain C.F. Schiller OAM CSM 2014-2015
Air Commodore K. Dunne 2016-2020

Vice-President—Defence Health

Rear Admiral G.S. Shirtley AM RFD RANR, 2011
Major General J.V. Rosenfeld AM OBE CStJ, 2012-2020

Editor: *Citizen Soldier*

Lieutenant Colonel B. Nebenzahl, 1988-1992
Lieutenant Colonel R.S.J. Thirwell RFD ED, 1993

Editor: *The Australian Reservist*

Lieutenant Colonel R. Sealey RFD ED, 1998-2007

Editor-in-Chief

Commander P. Hicks RFD RANR, 2008-2019

State Executives

New South Wales
President: Major General J.D. Keldie, 1992
Major General R.J. Sharp AO RFD ED, 1993
Major General R.J. Sharp, 1994
Major General R.J. Sharp, 1995
Major General W.E. Glenny AO RFD ED, 1996/1997
Squadron Leader C. Gilbertson, 1997/1998
Colonel G.W. Fleeton RFD, 1999-2020

Northern Territory
Unknown, 1993 1997
Lieutenant Colonel J. Dunn CSM RFD, 1998-2006
Colonel J. Dunn OAM CSM RFD, 2007-2019
2020

Queensland
1993
Lieutenant Colonel G.L. Hulse, RFD ED, 1994-1999
Brigadier P. Rule CSM, 2000-2004
Colonel G. Bulow RFD, 2005-2008
Major K. Lynch RFD 2009-2010
Lieutenant Colonel J. Cotton RFD, 2011-2016
Colonel C. Cuneen MSt.J, 2017
Colonel R. Olive AM RFD, 2018-2019
2020

South Australia, 1993
Lieutenant Colonel M.E. Alexander RFD ED, 1994-2009
Major General N. Wilson AM RFD, 2010-2015
Brigadier R.N. Atkinson AM RFD, 2016-2019
2020

Tasmania 1993
Lieutenant Colonel D.M. Wyatt RFD, 1994-1997

Patron: Sir Guy Green, AC KBE
Colonel J. Stewart MC, 1998-1999
Colonel D. Townsend RFD, 2000-2010
Colonel S. Carey RFD, 2011-2019
2020

Victoria
Colonel J. Sandow, 1993
Colonel W.M. Vincent RFD ED, 1994
Colonel W.M. Vincent RFD ED, 1995
Brigadier R.S.P. Amos RFD ED, 1996
Brigadier A.J. McGilliard RFD ED, 1997-2008
Major General G.H. Garde AO RFD QC, 2008-2011
Brigadier P.D. Alkemade RFD, 2012-2019
2020

Western Australia
1993
Lieutenant Colonel P. Farrell RFD ED, 1994-1995
Lieutenant Colonel P.G. Winstanley RFD, 1996-2000
Patron: Brigadier R.A. Lawler
Lieutenant Colonel R.G. Cook, AM RFD, 2001-2018
Colonel M. Page RFD, 2019-2020

SELECT BIBLIOGRAPHY

Archives and Reports

National Archives of Australia.

National Library of Australia and State Libraries.

Hansard, House of Representatives.
  –Parliamentary Debates.

Hansard, Senate.
  –Review of Australian Defence Capabilities, Report and Ministerial Statement, 3 June 1986.

Australian National Audit Office
  –Performance Audit Report – Army Reserves Forces, 8 May 2009.

Australian Defence Force Posture Review (A. Hawke and R. Smith), Department of Defence, 30 March 2012.

Dibb Report – *Review of Australia's Defence Capabilities* (P. Dibb), The Parliament of the Commonwealth of Australia, 1986.

Dunn/Sanderson Report –*Report on the Force Structure and Tasks of the Army Reserve* (B. Dunn and J. Sanderson), Army Reserve Review Committee, Oct 1986.

Joint Standing Committee on Foreign Affairs, Defence and Trade, *From Phantom to Force: Towards a More Efficient and Effective Army,* August 2000.

Millar Report – *Committee of Inquiry into the Citizen Military Forces* (T. B. Millar, Chair), AGPS, March 1974.

Tange Report–Australian defence: report on the reorganisation of the defence group of departments (A. Tange): presented to the Minister for Defence, Department of Defence, November 1973.

The Defence Reserves Association archives 1970-2020
 –Hardacre, D. 'The Early Years', 1996.
 –Glenny, W. 'The CMFA', unp.ms., 2020.

Books

Baker, K. *Paul Cullen: Citizen Soldier*, Rosenberg, 2005.

Carmichael, M. *In Kilted Company: A Story of Part-time Soldiering*, Melbourne Books, 2007.

Chapman, I. *Sydney University Regiment – the first 80 years*, I. Chapman, 1996.

Grey, J. *A Soldier's Soldier – A Biography of Lieutenant-General Sir Thomas Daly*, CUP, 2013.

Hurst, D. *The Part-Timers: A History of the RAAF Reserves 1948-1998*, ADF Journal, 1999.

Leckie, N. *Country Victoria's Own*, AMHP, 2008.

McCarthy, D. *The Once and Future Army*, OUP, 2003.

Palazzo, A. *Defenders of Australia – The Third Australia Division*, AHU Canberra, 2002.

Ryan, M.J. *A Most Unusual Regiment – A History of the Melbourne University Regiment*, AHMP, 2008.

Stockings, C. *The Torch and the Sword*, UNSWP, 2007.

Wilkins, J. M. *Naval Reserve 1859-2019*, Melbourne, 2019.

Articles and Papers

Armstrong, M. 'Pursuing the Total Force: Strategic Guidance for the Australian Defence Force Reserves in Defence White Papers since 1976', in *Security Challenges*, Vol.16, No.2. Special Issue, 2020.

Bevis A. 'New Reserve Directions' in *DRA NSW Newsletter*, January 1996.

Cullen P. A. 'Report Wrong on Two Vital Points', *PDR Charter Issue* 1975.

Cullen P.A. 'One Army', Blamey Oration, at RUSI NSW, 13 February 1991.

Day P.J. 'The Defence Force and the Community – A Tragedy in Ten Chapters', *RUSI NSW Journal*, article submitted 3 April 1991.

Glenny, W. 'The Defence Reserve and Its Role in the Defence of Australia', in *DRA NSW Newsletter*, April 1996.

–'An Army for the Twenty First Century—Fact or Fiction, Perceptions and Reality' in *DRA NSW Newsletter*, July 1997.

Lilley A.B. 'Sydney University Regiment: a description of the insignia worn from 1900–1973 by military units at the University of Sydney, together with information on honorary colonels and commanding officers', in *Sabretache*, Military Historical Society of Australia, 1974.

McCarthy D. 'Becoming the 3rd XV: The Citizen Military Forces and the Vietnam War', in 2002 Chief of Army's Military History Conference, AHU 2002.

Smith H. and Jans N. 'Stepping Up: Part Time Forces and ADF Capability', in *Strategic Insights*, No.44, ASPI, November 2008.

Tarnsitt G. 'The Army Reserve preparing for a Total Force' in *Australian Defence Force Journal*, No. 123, Sep/Oct 1997.

Magazines, Journals and Pamphlets

Australian Army Journal

Australian Reservist

Citizen Soldier

Defender

Lancer's Despatch

Newspapers and Internet

Advertiser

Australian Financial Review

Canberra Times

Courier Mail

Herald Sun

Monitor

News Weekly

Daily Telegraph

Telegraph

Weekend Australian

Websites

https://en.wikipedia.org/wiki/Australian_Army_Reserve

https://www.aph.gov.au/About_Parliament/Parliamentary_Departments/Parliamentary Library/pubs/ rp/ rp1516/ DefendAust/2013

http://adb.anu.edu.au/biography/callinan-sir-bernard-james-28251

https://www.globalfirepower.com/country-military-strength-detail.asp?country_id=australia

https://en.wikipedia.org/wiki/Defending_Australia_in_the_Asia_Pacific_Century:_Force_2030

https://en.wikipedia.org/wiki/Frank_Hassett

https://en.wikipedia.org/wiki/Green,Kenneth_David

https://en.wikipedia.org/wiki/James_Killen

https://en.wikipedia.org/wiki/Lance_Barnard

http://adb.anu.edu.au/biography/norris-sir-john-gerald-14998.

http://adb.anu.edu.au/biography/risson-sir-robert-joseph-17250.

Interviews 2020

Brian Nebenzahl,
Paul Irving,
Brian Smith,
Neil James,
Colin Gilbertson,
Neil Wilson,
Graham Fleeton,
Robert Atkinson,
Greg Garde,
Stefan Landherr,
Jim Barry,
Warren Glenny,
Joseph Lukaitis.

NOTES

1 Glenny, G. 'The CMFA', unp.ms. DRA archives 2020.

2 McCarthy, D. *The Once and Future Army*, OUP, 2003, p210. Cullen would later be AO and AC. Dayton stated that the CMFA was created in November 1969, but the first formal meeting was held in March 1970.

3 Glenny, G. *op.cit.*

4 'The Early Years', prepared by WO David Hardacre, DRA Victoria, 1996. The three others were not named. Letters of encouragement were received from Bruce White, Secretary of the Army Department and 'Tom', probably the CGS, Lieutenant General Sir Thomas Joseph Daly, KBE, CB, DSO.

5 McCarthy, D. *op.cit.* p215.

6 Sometimes the 'establishment' was a reference to Cullen himself. See Grey, J. *A Soldier's Soldier—A Biography of Lieutenant-General Sir Thomas Daly*, CUP, 2013, p127.

7 At the time Dart was also the manager of a large pharmaceutical distribution company and was able to travel across Australia at its cost, enabling him to keep up with Paul Cullen on his constant travel schedule. Interview with Brigadier Dart, OBE RFD ED, 8 May 2020.

8 Letter to potential members from Major General Cullen, 17 August 1970.

9 'The Early Years', prepared by WO David Hardacre, c.1996. Major General McDonald won his MC as a captain in New Guinea in 1945.

10 Major J R W Crook remains unidentified.

11 Glenny, G. *op.cit.* p2. McCarthy, D. *op.cit.* p211 also notes Major General Sir Frederick Galleghan as another Life Member at that time. Major Generals Simpson (1960-62), MacArthur-Onslow (1958-59), Risson (1957-58), Dougherty (1953-54) and Windeyer (1950-52) had all been CMF Members on the Military Board. Dougherty joined the CMF before WWII and saw service in the Middle east and New Guinea. He later became the first NSW Director of Civil Defence. He died in 1998.

12 The Royal Australian Navy Reserves, after a four year hiatus, had been re-activated on 1 January 1950 with the advent of the Korean War. By 1973 the RANR only held 600 'active ' reservists of all ranks against an official RANR Complement of 600 Officers and 2000 Other Ranks. See

Wilkins, J.M. *Naval Reserve, 1859-2019*, Robel, 2019, pp96, 105.

13 Paul Cullen 'Report to the Members of the [CMFA]', 1972.

14 CMFA National Council Accounts—Balance Sheets June 1972-June 1977.

15 Grey, *op.cit.* p125.

16 McCarthy, *op.cit.* p211.

17 *Citizen Soldier*, January 1972, p1.

18 Palazzo, A. *Defenders of Australia—The Third Australia Division*, AHU Canberra, 2002, p178. Compulsory military training post WWII was reintroduced in 1951. Eighteen-year-old men were required to undertake 176 days of military training—those who elected to undertake their training in the army could break up their training requirements into two periods, 98 days in the Australian Regular Army and 78 days in the Citizen Military Forces. In 1959 the scheme was abolished. https://www.awm.gov.au/articles/encyclopedia/conscription/vietnam

19 *Citizen Soldier*, January 1972, p1.

20 McCarthy, D. 'Becoming the 3rd XV: The Citizen Military Forces and the Vietnam War', *2002 Chief of Army's Military History Conference*, AHU 2002, p268.

21 *Citizen Soldier*, January 1972, p3.

22 McCarthy, *op.cit.* pp152-155. Cullen was successful in having CMF officers being attached to operational units in Vietnam for short periods—over 500 would gain some experience in that way.

23 Grey, *op.cit.* p127.

24 *Citizen Soldier*, *op.cit.* p1.

25 Letter, Major General Vickery to Major General Cullen, 26 August 1970, Series A6829, National Archives of Australia (NAA).

26 Murchison resigned as president of the CMFA NSW branch when he was appointed to the Military Board as CMF Member in 1973. He later became, after retirement, the CMFA president.

27 Letter, Major General Vickery to Major General Cullen, 26 August 1970, *op.cit.*

28 *Sydney Telegraph*, 14 August 1970.

29 Baker, K. *Paul Cullen: Citizen Soldier*, Rosenberg, 2005, p184, quoting a Dayton McCarthy interview with Maitland 22 October 1996. Maitland was a later president of the CMFA and AO.

30 Glenny was a later CMFA President. Regular officer Major General Francis George 'Frank' Hassett CB, DSO, LVO and later CGS and from 1976, Chief of the Defence Force Staff, (later knighted KBE, CB). https://en.wikipedia.org/wiki/Frank_Hassett

31 Glenny, *op.cit.* p2; https://en.wikipedia.org/wiki/Frank_Hassett; McCarthy, D. *op.cit.* p214.

32 ibid.

33 McCarthy, *op.cit.* pp85-125.

34 Grey, *op.cit.* p125. The Pentropic Division structure was disbanded in 1964

35 Palazzo, *op.cit.* p170.

36 McCarthy, *op.cit.* p165.

37 Glenny, *op.cit.* p2. Millar was a former regular army soldier 1943-1950). Four of his committee were current or former Reserve officers; only one was regular army.

38 *Citizen Soldier*, October 1973, p1. Lance Herbert Barnard AO (1919 – 1997) was a 9th Division veteran in the Middle East, including the battle of El Alamein. He was a MHR from 1954 to 1975 (Labor), and served as the party's deputy leader from 1967 to 1974. He served as Minister for Defence from 5 December 1972 to 6 June 1975. Post-politics, from 1981 to 1985, Barnard was director of the Office of Australian War Graves. https://en.wikipedia.org/wiki/Lance_Barnard

39 *Citizen Soldier*, October 1973, p5.

40 Glenny, *op.cit.* p2.

41 *Citizen Soldier*, *op.cit.* p5.

42 Committee of Inquiry into the CMF Report March 1974, AGPS 1974, p127.

43 'Report Wrong on Two Vital Points', Major General Cullen, PDR Charter Issue 1975, p17.

44 https://en.wikipedia.org/wiki/Australian_Army_Reserve.

45 Palazzo, A. *op.cit.* p188.

46 ibid. p189.

47 McCarthy, *op.cit.* p185.

48 ibid.

49 *Citizen Soldier*, June 1975, p2.

50 ibid.

51 ibid. Following the Millar Report, Murchison had been appointed as the first Inspector General of the Army Reserve in addition to his appointment as CMF Member. On his retirement in October 1974, he was replaced by Major General J M L Macdonald AO MBE RFD ED, as the last CMF Member.

52 Colonel T J Crawford (fnu) was a former CO of 19 Battalion (Bushmen's Rifles), RNSWR 1967-1972. He was a CMF Observer in the Korean War in 1952.

53 *Citizen Soldier*, December 1976, npn. Later Colonel Dunn OAM CSM RFD would return as president of the Defence Reserves Association (DRA) NT branch.

54 It is not known whether this conference took place.

55 Killen—later Sir Denis James Killen AC, KCMG—had served with the RAAF during World War II. He was a MHR from 1955 to 1983, and served as Vice-President of the Executive Council, Minister for Defence (1975-1982) and Minister for the Navy during his parliamentary career. https://en.wikipedia.org/wiki/James_Killen

56 *Citizen Soldier*, June 1975, p2. Issue of polyester uniforms and equal pay to regulars were implemented soon after.

57 Brigadier Grant was commander 3rd Division Field Force group 1976-1977; he was replaced by

Brigadier Jim Barry over 1977-1980. Major General Poke, AO RFD ED retired in 1980. He was CMFA Victorian vice-president in 1981. Lieutenant Colonel Bullard later became CMFA president in Victoria.

58 *Citizen Soldier*, December 1976, npn.

59 ibid.

60 *Citizen Soldier*, December 1976, npn.

61 Glenny, *op.cit*. p2.

62 Minutes, CMFA Vic, 3 March 1977.

63 Green was at the time Secretary of the Premier's Department in the Victorian Government. See his extensive biography at F. J. Kendall, 'Green, Kenneth David (Ken) (1917–1987)', Australian Dictionary of Biography, National Centre of Biography, Australian National University, http://adb.anu.edu.au/biography/green-kenneth-david-ken-12562/text22615, published first in hardcopy 2007, accessed online 21 August 2020.

64 Noted contributors at the discussion according to 'The Early Years', *op.cit*. included Brigadiers Eason, Sir Charles Chambers Fowell Spry CBE, DSO, Colwill and Sir William Hall. The meeting included five major generals, 14 brigadiers, 17 colonels and five lieutenant colonels. Spry, a Duntroon graduate and one time adjutant of the Sydney University regiment, had retired from the Army in 1954, and from 1950 to 1970 he was the second Director-General of Security, the head of the Australian Security Intelligence Organisation (ASIO). Brigadier Keith Royce Colwill CBE (later OAM) was also a Regular officer. He and Brigadier Sir William Hall KBE DSO ED were both Freemasons (as was Colonel, later Major General Frank Poke, who was at the meeting). Brigadier Richard Thomas 'Dick' Eason MC ED was named in 1972 as a member of the CMF Association national council representing Victoria so at least one of those at the meeting of July 1970 kept Cullen's flag flying in Victoria in the coming years. Another brigadier at the meeting, Brigadier Ian H Lowen OBE ED, was later noted as the Victorian 'president' on the 1975 CMFA national council (he was later to become the AResA president), while Victorians, Lieutenant General Sir Edmund Herring KCMG, KBE, DSO, MC, ED and Brigadier Richard Thomas 'Dick' Eason MC ED (also a Freemason) were one of the first CMFA life members and councillors respectively at the start of the association.

65 'The Early Years', *op.cit*. The Steering Committee consisted of Major Generals Fraser, Green and McDonald, Brigadiers Brown and Sir William Hall, Colonel Stewart Austin Embling and Brigadier Colwill, who was the Chief of Staff at Southern Command.

66 Annual Report, CMFA Vic, September 1976-December 1977, in Newsletter, DRA Vic, December 2006-January 2007.

67 Minutes CMFA Vic 17 November 1977. The meeting of 25 August 1971 included four major generals, nine brigadiers, 14 colonels, and seven lieutenant colonels. 'The Early Years', *op.cit*.

68 Hurst, D. *The Part-Timers: A History of the RAAF Reserves 1948-1998* ADFJ, 1999, pp42-48.

69 Minutes, CMFA Vic, 8 March 1977.

70 Minutes, CMFA Vic, 20 October 1977.

71 The Association engagement with CESRF, later the Defence Reserves Support Committee, has continued throughout the Association's history.

72 Minutes, CMFA Vic, 13 April 1978.

73 Hansard, House of Representatives, 24 October 1978, p2171.

74 Lieutenant General Sir Edmund Frederick Herring KCMG, KBE, DSO, MC, KStJ, ED, KC had retired as Chief Justice of Victoria in 1964; CMDR Sir Neville Stewart Pixley CMG, MBE, VD, KtON, RANR was a former ADC to the Queen. ACM Sir Frederick Rudolph Scherger KBE, CB, DSO, AFC, retired as head of the RAAF and as chairman of the Chiefs of Staff Committee in 1966.

75 The CMFA was in fact correct—the ED and EM were in fact Australian awards approved by the Queen, and so argued for 'the retention of these traditional awards'—it seems the nuance was lost on the Government which clearly wanted a clean break from the past.

76 Minutes, CMFA Vic, 6 June 1978.

77 Hurst, D. *op.cit.* p55.

78 Coldham was a former RAAF officer who had seen active service in Bomber Command in WWII, and who was awarded a DFC and bar.

79 *Parliamentary Debates*, House of Representatives, Vol.6122 October 1968, p2217.

80 Minutes, CMFA Vic, 13 April 1978.

81 It is not known how often, other than for the AGM, the Executive of the National Council met or whether it met within the context of the NSW branch at this time.

82 Major General G Maitland was later the first Reserve officer to be CRES (1979-1982) and a member of the Chief of Army's Advisory Council.

83 Minutes, CMFA Vic, 4 September 1978.

84 ibid. The ADA was founded in Perth in mid-1975

85 Minutes, CMFA Vic, 4 September 1978 ibid.

86 ibid.

87 Annual Report to Members of the [CMFA], 1 November 1979.

88 Major General J M L Macdonald was the last CMF member of the Military Board. He had served with distinction in the British Army in WWII and had been wounded in action. He migrated to Australia after the war and continued his career in the CMF. He was Aide-De-Camp to the Queen in 1971 and was Commander 2 Division in 1973. Later he was Hon. Colonel to 1/15 Royal NSW Lancers. DRA NSW Newsletter 1/96, February 1996.

89 Minutes, CMFA Vic, 20 September 1979.

90 ibid.

91 Glenny, *op.cit.* p3. RCSC were established in each State. They were sometimes referred to as 'Retired Colonels Social Clubs', although by whom can only be guessed at.

92 There were doubts whether this could be sustained. According to one calculation, the processing of 500 recruits a week for seven weeks would exhaust the supply of uniforms and necessities.

93 Minutes, CMFA Vic, 15 May 1980.

94 Minutes, CMFA Vic, 15 May 1980.

95 ibid.

96 In early 1981 the Active Citizens Air Force (ACAF) was renamed the RAAF Active Reserve (RAAFAR). About this time, the RAAFAR began accepting female members, while aircrew members were also accepted for continuation of their flying duties if they left the PAF before reaching mandatory retirement age. Increased use was also made of RAAFAR members to provide relief manning to the PAF. Notes by Squadron Leader Gilbertson, email to author, 30 July 2020.

97 Minutes, CMFA Vic, 11 September 1980.

98 Minutes, CMFA Vic, 11 September 1980.

99 Minutes, CMFA Vic, 4 December 1980.

100 Minutes, CMFA Vic, 4 December 1980.

101 ibid.

102 Minutes, CMFA Vic AGM, 23 April 1981.

103 Minutes CMFA Vic, 16 July 1981.

104 Minutes, CMFA Vic, 3 September 1981.

105 Lieutenant General Sir Mervyn Francis Brogan KBE CB, the former CGS, retired in early 1975 and was Honorary Colonel of the UNSWR from 1975-1980.

106 Minutes, CMFA Vic, 5 November 1981. The Constitution sub-committee met again in Sydney in February 1982 and further drafts and re-drafts continued right up to the August 1983 CMFA AGM.

107 Interview Major General W Glenny 10 March 2020.

108 Minutes, CMFA Vic, 5 November 1981.

109 Major J R W Crook remains unidentified.

110 Whiting replaced Brigadier John Anthony Springhall who had retired. Whiting was Director of the Queensland State Emergency Services.

111 *Citizen Soldier*, April 1982, p7. Lieutenant General Sir Edmond Herring KCMG KBE DSO MC ED remained on the books as a Life Member but passed away in January 1982. During World War II, Herring commanded the 6th Division Artillery in the Western Desert Campaign and the Battle of Greece. In 1942, as a corps commander, he commanded the land forces in the Kokoda Track campaign. In 1943, he directed operations in the Salamaua-Lae campaign and Finisterre Range campaign. Herring left his corps to become the longest-serving Chief Justice and Lieutenant Governor of Victoria, serving for three decades. He was also Honorary. Colonel of Melbourne University Regiment (MUR) for more than 30 years.

112 Kevin Newman AO, MHR (Liberal, Bass) was a former lieutenant colonel in the ARA who had served in Malaya and Vietnam. He was Administrative Services Minister 1980-83. Major

General Cullen was president of the NSW CESRF in 1981.

113 Minutes, CMFA Vic, 3 September 1981.

114 ibid.

115 *Citizen Soldier*, April 1982, p1.

116 Robert Cummin Katter, who served in Army in WWII was a Country Party MP who served as Minister for the Army in the McMahon Government from 1971 to 1972.

117 Minutes, CMFA Vic, 25 March 1982.

118 *Citizen Soldier*, April 1982, p1.

119 Minutes, CMFA Vic, 17 June 1982.

120 Minutes, CMFA Vic, 17 June 1982. Brigadier John Gilbert McKinna CMG CBE DSO LVO ED was an infantry officer during WWII and later served as Commissioner of the SA Police (1957–72).

121 Minutes, CMFA Vic, 17 June 1982.

122 Minutes, CMFA Vic, 17 June 1982.

123 Minutes, CMFA Vic, 17 June 1982.

124 Ian McCahon Sinclair AC served in 22 Squadron RAAF from 1950 to 1952, as part of the Citizen Air Force. A leader of the National Party, he was Minister for Defence in the Fraser Government in 1982-83.

125 Minutes, CMFA Vic, 17 June 1982.

126 Covering letter, Brigadier Lowen to G Scholes, *Submission on the Australian Army*, June 1983. The submission was then sent to the National Council, CRES etc.

127 The Defence Act would shortly be amended, changing CMF to Army Reserve.

128 Minutes, CMFA Vic, 15 September 1983.

129 Minutes, CMFA Vic, 5 April 1984.

130 In another example of looking for support from well-known individuals especially with political leaders, Lieutenant Jeff Kennett, a future State premier, joined the CMFA Victorian branch. Minutes, CMFA Vic Committee, 7 April 1983.

131 Minutes, CMFA Vic, 1 December 1983.

132 Ken Aldred MHR (Liberal) served in the ARes 1965-1971. He was MHR for Henty 1975-1980, Bruce 1983-1990, and Deakin 1990-1996. During his time as the MHR for Bruce, Aldred took a prominent role on defence issues, as chair of the Opposition Defence Committee.

133 Minutes, CMFA Vic, 5 April 1984.

134 *Hansard*, House of Representatives, No137 31 May 1984, pp2601-2602.

135 CMFA AGM Agenda, 13 August 1984.

136 Sandow had joined the CMF in 1948 in South Australia and was the first Commandant of the Reserve Staff and Command College in Victoria. He retired in 1990, joining the CMFA and

becoming president of the Victorian branch in 1992 and national president in 1994.

137 MacLeod was replaced by Major Ian Haskins in April 1985.

138 Minutes, CMFA Vic, 6 December 1984.

139 Minutes, CMFA Vic, 6 December 1984.

140 Minutes, CMFA Vic AGM, 20 September 1984. The proposal went to the RSL's National Executive in February 1985.

141 Report by President [Brigadier Lowen] 1984-1985, 20 August 1985, DRA Archives.

142 CMFA Submission to the DFRT, 17 May 1985.

143 Minutes, CMFA Vic, 11 April and 13 June 1985. Minister for Defence Kim Beazley commissioned Paul Dibb, an external consultant and former employee of the Department of Defence, to analyse Australia's defence planning and make recommendations for future developments.

144 Letter, Major General Cullen to Brigadier Lowen, 12 October 1984, DRA Archives.

145 Minutes, CMFA NSW, October 1984.

146 Letter, Brigadier Lowen to Brigadier Amos, 4 March 1985. In the Australia Day Honours List of 1985 only two ARes officers received awards compared to 21 in the ARA.

147 Minutes, CMFA NSW, 29 November 1984 and CMFA NSW Paper, 'Promotion Problems in the Army Reserve', Lieutenant Colonel W B Molloy RFD ED, 11 February 1985.

148 Only the June 1985 Newsletter survives in DRAC Archives.

149 CMFA National Council Account Auditor's Report, 26 August 1985. Lieutenant Colonel A S Furze RFD ED, a chartered accountant, was the auditor.

150 CMFA Submission 'The Australian Army's Rapid Deployment Capability' to the Standing Committee on Foreign Affairs and Defence, 28 May 1985, DRA Archives.

151 Letter, Minister Beazley to Brigadier Lowen, 17 January 1985.

152 CMFA Newsletter, June 1985.

153 *Hansard*, 22 May 1985, pp2900-2909.

154 CMFA Newsletter, June 1985.

155 Duffy was the Minister for Communications. Major General James Edward Barry MBE RFD ED was a former Olympics administrator—and future DRA president.

156 CMFA Notice of Meeting 29 August 1985, DRA Archives.

157 Minutes, CMFA Vic AGM, 17 October 1985.

158 CMFA Newsletter February 1986.

159 Minutes, CMFA NSW, October 1984.

160 Minutes, CMFA AGM, 29 August 1985.

161 CMFA Newsletter February 1986.

162 Minutes, CMFA AGM, 29 August 1985. Harradine later raised the question of housing loans in a

debate on the Appropriation Bill on 27 November 1985 but was rebuffed by the Government, which 'had no plans to introduce such a scheme'. CMFA Newsletter, February 1986.

163 The CMFA position was that any ARes members who had served for 14 days in South Vietnam (SVN) should be awarded the Vietnam service medal. This was rejected in January 1987 on the basis that the ARes officers were attached and not posted to the units they visited. They were allowed the Returned from Active Service badge. Letter to Brigadier Lowen from Kim Beazley, 20 January 1987, DRA Archives.

164 Minutes, CMFA Vic AGM, 17 October 1985. Hodges was later appointed Commander 5 MD and he was replaced in 1987 as DCRES by Brigadier Brian Edwards. CMFA Newsletter November 1986.

165 Report by President [Brigadier Lowen] 1984-1985, 20 August 1985.

166 Minutes, CMFA Vic AGM, 17 October 1985.

167 By April 1986 members in the CMFA NSW branch had dropped to 36 and in June the branch only had $365 in the bank. Minutes, CMFA NSW 10 April 1986 and CMFA NSW Financial Report 26 June 1986.

168 Report, CMFA President [Brigadier Lowen] 1984-1985, 20 August 1985.

169 CMFA Newsletter November 1985.

170 CMFA Newsletter July 1986.

171 CMFA Newsletter February 1986.

172 CMFA Newsletter, February 1986.

173 Minutes, CMFA NSW, 27 February 1986.

174 Newsletter, CMFA NSW, June 1986.

175 The Defence White Paper would be released in March 1987.

176 CMFA Newsletter, June 1986.

177 https://en.wikipedia.org/wiki/Australian_Army_Reserve

178 Hansard, 'Review of Australian Defence Capabilities, Report and Ministerial Statement', 3 June 1986, p4419.

179 Speech Notes, Major General O'Donnell for Brigadier Amos, 30 June 1986.

180 CMFA Newsletter July 1986.

181 ibid.

182 Minutes, CMFA AGM, 28 August 1986.

183 Minutes, CMFA AGM, 28 August 1986.

184 The RSL, AFFA and the Australian Medical Association had all been granted permission to appear before the Tribunal. The CMFA sought the same permission but generally relied on the Defence Force Advocate to present its case. Minute, Chief Personnel—Army to CRES, 27 May 1986.

185 Later Major General (and Chief Justice) Garde was then CO of 4/19 Prince of Wales Light Horse, former CO of Monash University Regiment and a barrister and solicitor; Lieutenant Colonel

Furze was head of 3 Div Pay Corps and a chartered accountant with Deloitte, Haskins & Sells; and Dr Jessup was an industrial lawyer (Harvard) and former sergeant in MUR.

186 Letter to Lieutenant Colonels Staziker, Broadhead and Keefe from Brigadier Amos, 12 December 1986.

187 Letters, Brigadier Amos to Brigadier Lowen, 26 September 1986 and reply, 10 October 1986.

188 Letter to Brigadier Amos from Lieutenant Colonel MacLeod, 29 May 1987.

189 CMFA NSW Report on the Army Reserve Pay and Conditions of Service, 9 March 1987.

190 CMFA President's Report, 1986.

191 ibid.

192 ibid.

193 ARESA Newsletter November 1986, incorporating Minutes of the ARESA AGM.

194 Later Lieutenant General Sanderson AC and Governor of WA. Later Major General Nunn AO RFD ED and Commander 3 Division.

195 Later Colonel Sibree AM MG. In 1987 appointed Commandant, Infantry Centre, Singleton. Colonel Johnson CSC AAM RFD ED was, in 1987, a member of the Reserve Staff Group and seconded to the ARRC Implementation Team. He had spent 14 months in Antarctica as Officer-in-Charge of Casey Station with the Department of Science in 1981.

196 Gration succeeded Lieutenant General Bennett as CDFS in April 1987.

197 *Report on the Force Structure and Tasks of the Army Reserve*, Army Reserve Review Committee, Oct 1986.

198 AResA National Newsletter incl. 'The Future of the Army Reserve', March 1987.

199 Minutes, AResA NSW 14 May 1987.

200 Minutes, AResA AGM, 24 September 1987. The loss of a battalion in the Newcastle area was especially objectionable.

201 AResA National Newsletter, March 1987. Harrison had joined the Queensland University Regiment (QUR) in 1948 and later became the first recruit to also become Commanding Officer.

202 Address, 'Evolution of an Army', Major General Cooke to Imperial Service Club, Sydney 26 June 1987.

203 Minutes, AResA NSW, 17 September 1987.

204 Minutes, AResA AGM, 20 August 1987.

205 ibid.

206 Minutes, AResA AGM, 20 August 1987.

207 ibid.

208 Cullen, and former AResA president Major General Macdonald, continued to represent the Association at CESRF meetings.

209 Lieutenant Colonel Staziker was CO of 7 Field Regiment. Lieutenant Colonel Keefe was a chartered accountant. A former CO of 7 Field Regiment himself, he rose to become Commander

8 Brigade. Lieutenant Colonel Broadhead was a Reserve Bank officer and at the time CO of 23 Field Regiment.

210 In one of those small changes to ARes payments, modern times reached out with integration that year to enable ARes pay by direct credit rather than by cheque!

211 Email, Major General Garde to author, 10 June 2020.

212 Minutes, Extraordinary General Meeting AResA Vic 25 February 1988.

213 Minutes AResA Vic 13 October 1988. At this stage Victoria branch membership was 240 but only 60 were financial. In February 1988 David Quick QC was replaced by Michael Black QC on the Pay & Conditions Tribunal.

214 AResA National Newsletter, May 1987 and the *Weekend Australian*, 22-23 April 1989, p5.

215 Major General Fay, who joined the CMF after National Service in 1951, was an industrial chemist and later general manager within the food industry. He retired from ACGS-Res in March 1990, became AResA president 1991-1994 and remained involved with the Defence Reserves Support Committee.

216 AResA National Newsletter, June 1988.

217 Minutes, AResA NSW, 15 November 1988 and 12 January 1989.

218 Minutes AResA Vic 11 February 1988.

219 'Army Reserve Future Directions' (Dart, Parsonage and Willis), March 1988. Brigadier Dart, first secretary to the CMFA had been Commanding officer of Sydney University Regiment 1967-1970. Brigadier Willis, a solicitor, was a Liberal member of the New South Wales Legislative Council from 1970 to 1999. He was president of the Imperial Service Club 1972-1975, and commanded UNSWR 1972—1975. Brigadier Parsonage was the former Brigadier (Plans) at Headquarters Field Force Command. He had been UNSWR Commanding Officer 1965-1968 and later became UNSWR Honorary Colonel 1991-1997.

220 'Army Reserve Future Directions'(Dart, Parsonage and Willis), March 1988.

221 Email Major General Garde to author, 9 June 2020.

222 Minutes AResA Vic 13 October 1988.

223 Newsletter, AResA NSW, March 1989.

224 Ryan, M.J. *A Most Unusual Regiment*, AHMP, 2008, p139.

225 Report, *op.cit.* pp4-13. The appointment of regular Colonels to replace ARes brigadiers as commanders of 4 and 5 Brigades in 1989 was a case in point.

226 Minutes, AResA Vic, 13 October 1988.

227 AResA Submission to Parliamentary Defence Sub-Committee, 29 June 1989.

228 Keynote Address, Major General Smethurst, CRES Conference, 31 March 1989.

229 *Hansard* (Senate), 8 June 1989, p3670.

230 Letter, Major General Cullen to PM Hawke, 10 November 1987.

231 Letter, Major General Cullen to PM Hawke, 26 September 1988.

232 Letter, Major General Cullen to GEN Gration, 25 January 1989.

233 Letter, Major General Cullen to Minister Simmons, 28 June 1989.

234 Letter, Major General Cullen to Lieutenant General O'Donnell, 21 July 1989

235 Letter, Major General Cullen to D W Simmons, 21 July 1989.

236 Letter, D W Simmons to Major General Cullen, 14 August 1989.

237 Minutes, AResA AGM 7 October 1989.

238 Letter, Lieutenant General O'Donnell to Major General Cullen, 14 September 1989.

239 Minutes AResA Vic 13 October 1988.

240 Submission, AResA Vic, 5 May 1989.

241 Minutes, AResA NSW 13 May 1988 and 21 July 1988.

242 Minutes, AResA NSW, 12 January 1989.

243 Minutes, AResA NSW, 9 March 1989.

244 Minutes, AResA NSW 1 June 1989.

245 Minutes, AResA NSW, 6 September 1989. Email, to Colonel Fleeton from Squadron leader Gilbertson 10 February 2020. Gilbertson was the TRA treasurer.

246 Letter, Brigadier Amos to Alan Stephens, Secretary Defence Sub-Committee, JPFADT, and attached submissions, 1 June 1989. Major General Fay as CRES also made a lengthy submission as did Major Generals Cooke, Keldie and Murray, among others.

247 The Association submission was finally made in November 1989 after Brigadier Amos was unable to present at the Sub-Committee public hearings due to scheduling issues.

248 Letter, Brigadier Amos to Alan Stephens, *op.cit.* 1 June 1989.

249 Minutes, AResA NSW 1 June 1989. Broadbent, like Cullen, had served in the Middle East and New Guinea in WWII. He retired in 1966 after commanding 2 Division.

250 Submission No 73, 'Report on The Reserve Component of the Australian Defence Force…' by Ian Andrew Mott, Von Nida Dews and Sachs Pty Ltd, received 1 July 1989.

251 AResA National Newsletter 90-1, February 1990.

252 Letter covering Submission, Major General Fay to Brigadier Amos 16 June 1989. Fay's submission ran to 25 pages.

253 Address to AResA NSW by Major General Fay, 13 September 1989.

254 Letter covering Submission, Major General Cullen to Peter Gibson MC, Secretary Joint Parliamentary Committee on Foreign Affairs, Defence and Trade (Defence Sub-Committee), 3 October 1989.

255 ibid.

256 Letter, Major General Cullen to Lieutenant General O'Donnell, 3 October 1989.

257 Minutes, AResA AGM, 7 October 1989.

258 In other appointments, Major General Nunn became ACRES and Major General Keldie became

Commander 2 Division. Keldie was a Regular officer who had won his MC during Vietnam service.

259 AResA National Newsletter December 1989.

260 AResA President Report, August 1990.

261 Minutes, AResA AGM, 7 October 1989.

262 Minutes, AResA AGM, 7 October 1989.

263 Letter Colonel O'Dempsey to Brigadier Amos, 27 February 1991.

264 Minutes, AResA AGM, 7 October 1989.

265 Brigadier Standish had recently retired as Commander 3 Training Group and would later become Colonel Commandant of the Royal Regiment of Australian Artillery; at the time he was also finance controller for a large multinational. Lieutenant Colonel Finnane, a Sydney barrister, later Colonel in the Australian Army Legal Corps.

266 AResA National Newsletter December 1990.

267 Major General Murray, a prominent Sydney barrister, was CRES in 1982 and later, on retirement in 1985, Hon. Colonel of Sydney University Regiment.

268 AResA National Newsletter 90-1, February 1990. Major General Keldie was Commander 2 Division.

269 Senator Ray was Minister for Defence 1990-1996. Lindsay was a former major in the Legal Corps.

270 AResA National Newsletter August 1990.

271 AResA National Newsletter 90-2 April 1990.

272 Letter Brigadier Amos to State Presidents AresA, covering Future Directions Paper, 5 March 1990 and Newsletter, AResA NSW branch 12 September 1990.

273 AResA National Newsletter June 1990. DFRDB—Defence Forces Retirement and Death Benefits (scheme).

274 Minute, Lieutenant Colonel Hallinan to AResA & NSW branch committees, 20 July 1990.

275 Minutes, AResA AGM, 25 August 1990.

276 AResA Draft, Comment on the Wrigley Paper, 11 October 1990.

277 Defence Press Release, 13 August 1990.

278 AResA Draft, *op.cit.* 11 October 1990.

279 Submission, AResA Vic, 8 October 1990.

280 Letter Major General Fay to Brigadier Amos, 9 October 1990.

281 AResA National Newsletter 90-4, December 1990.

282 Major General P J Day, AO, 'The Defence Force and the Community—A Tragedy in Ten Chapters', RUSI NSW Journal article submitted 3 April 1991.

283 Minutes, AResA AGM, 25 August 1990.

284 The Government agreed to a new MCS scheme in June 1992.

285 AResA National Newsletter 90-3, 7 September 1990.

286 Minute, Lieutenant Colonel Hallinan to AResA & NSW branch committees, 20 July 1990.

287 *Hansard* (Senate), 13 November 1990, pp4062-4069.

288 ibid.

289 Bilney was Minister for DS&P from 1990-1993.

290 AResA National Newsletter 90-4 December 1990. Peter Lindsay was not related to Eamon Lindsay, who had held Herbert for Labor.

291 AResA National Newsletter 90-4 December 1990.

292 Letter, Major General Vickery to PM Hawke, 14 January 1991.

293 AResA National Newsletter 90-4 December 1990.

294 The story of 25 Squadron is a good example of the journey of RAAF Reserve units. In 1960 the squadron ceased flying duties and switched to the ground support role. In 1989, flying operations resumed with the Macchi MB-326 as No. 25 Squadron assumed responsibility for jet introduction training and fleet support; this role ceased in 1998 and since then the squadron has been tasked with providing a pool of trained personnel to the Air Force. https://en.wikipedia.org/wiki/No._25_Squadron_RAAF

295 Blamey Oration, 'One Army', Major General Cullen at RUSI NSW 13 February 1991.

296 Letter, AResA President to members, 14 February 1991.

297 Minutes, AResA NSW, 6 June 1991 and Minutes AResA AGM, 31 August 1991.

298 Minutes, AResA AGM, 31 August 1991.

299 AResA AGM, SA Report, 1991.

300 AResA National Newsletter 91-2, June 1991. Senator Durack was at the time Deputy Leader of the Opposition.

301 Letter covering AResA submission, Brigadier Amos to Ms. J Towner, Sub-Committee Secretary, 18 June 1991.

302 Letter covering Victorian submission, Colonel Clendinnen to Brigadier Amos, 26 June 1991.

303 Minutes, AResA NSW AGM, 26 September 1991. In other changes affecting the Reserves, Major General Nunn became ACDF—Reserves advising the CDF on all Reserve matters and being responsible for CESRF as well. Major General Luttrell became AC-RES replacing Nunn.

304 AResA National Newsletter 91-3, 23 July 1991.

305 AResA National Newsletter 92-1, February 1992.

306 JCFAD&T media Release, 28 November 1991.

307 *Hansard* 19 August 1992, p271.

308 Newsletter, DRA Vic September 2006.

309 Letter to the Editor *Herald-Sun* from Colonel Sandow, [unpublished], 5 December 1991.

310 Letter to the Editor *Courier-Mail* from Colonel Sandow [unpublished], 5 December 1991.

311 Letter to the Editor from Colonel Sandow, *Weekend Australian*, 15-16 February 1992.

312 Stockings, C. *The Torch and the Sword*, UNSWP, 2007, pp205-207.

313 Minutes CMFA Vic, 1 September 1983 and 5 April 1984.

314 Minutes, CMFA Vic December 1984

315 Minutes, CMFA Vic 14 February 1985.

316 Stockings, *op.cit.* p212.

317 AResA National Newsletter, March 1987.

318 Stockings, *op.cit.* pp177-182.

319 Minutes, AResA NSW, 23 January 1992 and Minutes, DRA NSW, 30 April 1992.

320 The Citizen Air Force (CAF) existed prior to the Second World War and the first two CAF units, Nos 21 and 22 Squadrons were formed on 20 April 1936. Three further squadrons (Nos 23, 24 and 25) were subsequently formed and operated as combined PAF/CAF units during the war—most were disbanded in 1946 and 1948. Four squadrons (Nos 21, 22, 23 and 25) were reformed as CAF units in 1948, but 24 Squadron was not reformed until 1951. During that time, the CAF was involved in flying and ground training. In June 1960, when CAF flying operations were ceased and the squadrons converted to a non-flying role supporting RAAF activities in their region. In 22 Squadron's case, that was direct support to RAAF Base Richmond. Notes from Squadron Leader Gilbertson, email to author, 30 July 2020.

321 Senator MacGibbon was formerly Deputy Chair of the Joint Parliamentary Committee which had conducted the Inquiry into Defence Reserves in 1990; Major General Nunn (ACRES, HQADF); Major General Luttrell ACRES-A, Group Captain A R Barr, DRES-AF and Captain K Taylor RANR, were also speakers.

322 DRA Queensland, Report to National Council, August 1992.

323 DRA President's Report to AGM, 29 August 1992.

324 DRA President's Report to AGM, 29 August 1992.

325 DRA Newsletter, March 1993.

326 Letter, Major General Fay to DRA State presidents, 27 January 1993.

327 Email Squadron Leader Gilbertson to Colonel Fleeton, 10 February 2020.

328 The Government announced a broader committee to Review of Australian Honours and Awards in October 1994.

329 Letter, Major General Fay to Major General Nunn, 23 July 1993.

330 Letter, Vietnam Logistic Support Veterans Association WA to John Faulkner MP, 4 June 1993.

331 Army Office Minute A93-17517, Lieutenant General Grey to Senator Ray, July 1993.

332 DRA President's Newsletter, March 1994. The bill amending the SRCA was rejected in March 1994.

333 DRA Tasmania Information Brief No10, January 1997.

334 DRA Tasmania Annual Report, 4 October 1996.

335 DRA Vic President's Report, September 1993. Sandow gave no examples of those occasions.

336 DRA Constitution and Rules Draft, 23 July 1993.

337 DRA President's Newsletter, March 1994.

338 DRA President's Newsletter, March 1994.

339 DRA WA Newsletter, 'Reservist', Vol2 Issue 2, September 1994 and Minutes, DRA WA, 9 September 1994.

340 Later Commodore Taylor, AM, who was a Judge of the NSW District Court and Deputy Judge Advocate for the ADF for seven years.

341 DRA President's Newsletter, March 1994.

342 DRA President's Report, 23 October 1993.

343 Minute Covering submissions, Major General Fay to Major General Nunn, 7 February 1994.

344 DRA President's Newsletter, March 1994. The DFRT confirmed that submissions could be made from November 1993.

345 Letter, Major General Fay to Colonel Sandow, 24 December 1993.

346 Minutes, DRA Victoria, 9 March 1994. Peter Charlton, 'Will Reserve Fight', *Monitor*, 15 January 1994. Garde later became a major general AO, and also a Justice of the Supreme Court of Victoria.

347 Minutes, DRA Victoria, 9 February 1994.

348 Minute, 2 Div HQ (Glenny) to LHQ, 23172/93, 3 February 1993.

349 DRA President's Newsletter, March 1994.

350 Letter Major General Fay to Major General James, 4 March 1994. Colonel Sandow also represented DRA on the Kindred Organisations Committee (KOC).

351 Email Squadron Leader Gilbertson to Colonel Fleeton, 10 February 2020.

352 Paper, 'The Protection Required for the Civilian Interests of Reservists When Called Out', Major General Luttrell, March 1994.

353 Letter, Major General Fay to Major General Nunn, 7 February 1994.

354 Minutes, DRA Victoria 12 April 1995.

355 DRA President's Report, 19 May 1994.

356 Letters, Colonel Sandow to Major General Luttrell and Major General Golding, 19 July 1994.

357 Draft paper for CGSAC, 'Reserve Development Strategy', under cover Major General Cooke 29 August 1994.

358 Minutes, DRA Victoria, 12 May 1995.

359 DRA NSW Newsletter 2/94, October 1994.

360 DRA NSW Newsletter 2/94, October 1994.

361 Note, Major General Sharp to Colonel Sandow, 7 October 1994.

362 Later observers credited the RRes with producing a first-class cohort of junior leaders for the

Reserve.

363 Letter, Colonel Sandow to Peter Reith, 14 October 1994.

364 Minutes, DRA Tasmania, 2 September 1994. In May 1994, NSW branch noted that of 205 nominal members only three were from the RAAF and only nine from the RAN—DRA NSW membership Summary, 27 May 1994 and Minutes DRA NSW, 6 October 1994.

365 ANR Briefing by Commander P. Hardy, DRA Queensland, 8 February 1995.

366 ibid.

367 Minutes, DRA Victoria, 12 April 1995.

368 Letter, Colonel Sandow to Major General Glenny, 23 May 1996.

369 DRA NSW Newsletter 2/95, August 1995.

370 Minutes DRA Victoria, 14 June 1995.

371 Letter covering DRA Queensland Workshop Report, Lieutenant Colonel Hulse to Colonel Sandow, 24 July 1996.

372 DRA Tasmania Q3 Newsletter, 1995.

373 Annual Report, DRA Queensland, 3 October 1995. NSW branch was well established by then. In December it reported 159 ordinary and 73 'fully subscribed ('life') members. Minutes, DRA NSW, 14 December 1995.

374 DRA NSW Newsletter 1/96, February 1996.

375 'New Reserve Directions', Arch Bevis MP, 22 January 1996.

376 Facsimile, Colonel Sandow to Major General Cooke 24 January 1996.

377 DRA NSW Newsletter 2/96, August 1996.

378 Minutes, DRA Victoria, 14 March 1996.

379 'ADF in the Twenty-First Century (Personnel Policy Strategy Review—The Glenn Report)', pp183-189, 1995.

380 Minutes, DRA Victoria, April 1996.

381 'The Defence Reserve and Its Role in the Defence of Australia', Major General Glenny, 23 April 1996.

382 ibid.

383 Facsimile, Colonel Sandow to Lieutenant Colonel Hulse, 10 April 1996.

384 Minutes, DRA NSW, 13 June 1996.

385 DRA Policy Statement, 31 May 1996.

386 Minutes, DRA Tasmania, 16 May 1996.

387 Minutes DRA Victoria, 11 July 1996.

388 'The Defence Reserve and Its Role in the Defence of Australia', Major General Glenny, 23 April 1996.

389 *Australian*, 'Go slow for troops as budget blows out', 1 July 1996.

390 *Australian*, 'Army Reserve's role under surveillance, ready of not', 2 July 1996.

391 *Australian*, 'Myth of the Reservists', 5 July 1996.

392 ibid.

393 *Australian*, Letters to the Editor, 'Senior Reserve officers' attitude dismaying' 4 July 1996 and 'Myth of the Reservists', 5 July 1996.

394 Memo, COL Sandow to Major Generals Glenny and Cullen, 30 July 1996.

395 ibid.

396 Memo, Colonel Sandow to Major Generals Glenny and Cullen, 30 July 1996.

397 DRA NSW Newsletter 2/96, August 1996.

398 ibid.

399 DRA NSW Newsletter 2/96, August 1996.

400 DRA NSW Newsletter 2/96, August 1996.

401 ibid.

402 DRA NSW Newsletter 1/96, February 1996.

403 DRA NSW Newsletter 2/96, August 1996.

404 Lieutenant General Frank J Hickling AO CSC became Chief of Army (CA) from 24 June 1998. Lieutenant General Sir Reginald George Pollard, KCVO, KBE, CB, DSO, was CGS 1960-1961.

405 Letter, Major General Cullen to Ian McLachlan, 4 September 1996.

406 Letter, Major General Sharp to Major General Hickling 6 September 1996.

407 Letter, Peter Jennings to Major General Glenny, 23 September 1996.

408 Some branches were robust around this time—NSW branch reported 177 ordinary members and 76 fully subscribed (life) members for a total of 253, while Tasmania branch under Lieutenant Colonel Douglas M Wyatt RFD noted 290 members including 54 life members in newsletter in January 1997. Minutes, DRA NSW, 5 September 1996, and DRA Tasmania Information Brief No10, January 1997.

409 DRA NSW Newsletter 1/97, June 1997.

410 *Sydney Morning Herald*, 4 December 2014.

411 DRA NSW Newsletter 1/97, June 1997.

412 Press Release, Minister for Defence, Science and Personnel, 30 October 1996.

413 Navy required a Reservist to complete two years full time service to qualify as a seaman officer. Minutes, DRA Victoria, 11 July 1996.

414 Letter, Colonel Sandow to Ian McLachlan MP, 12 December 1996.

415 ibid.

416 Excerpt, *The Logbook*, p5, in DRA NSW Newsletter 1/98, April 1998.

417 Minutes DRA Victoria, 12 December 1996.

418 Minutes DRA Victoria, 12 December 1996.

419 Letter, Major General Cullen to John Howard, 20 December 1996.

420 ibid.

421 ibid.

422 RUSI president was Major General Maitland, a former AResA president.

423 DRA NSW Newsletter 1/97, June 1997.

424 Minutes, DRA NSW, 23 July 1998

425 Minutes, DRA NSW, 6 February 1997.

426 Letter, Colonel Sandow to Major General Glenny, 18 December 1996.

427 Email, Squadron leader Gilbertson to Colonel Fleeton, 10 February 2020. The NSW branch also had a senior Navy representative on the committee, Captain Edward D Hearne ED, RANR.

428 DRA President Update, 25 September 1997.

429 DRA President Update, 25 September 1997.

430 ibid.

431 Major General Glenny, 'An Army for the Twenty First Century—Fact or Fiction, Perceptions and Reality', July 1997.

432 ibid.

433 DRA NSW Newsletter 1/98, April 1998. The Tange Report or 'Australian Defence: Report on the Reorganisation of the Defence Group of Departments' created the ADF and the Australian Defence Organisation.

434 ibid.

435 ibid.

436 Minutes DRA AGM, 16 November 1997.

437 DRA NSW President Report, 29 September 1998.

438 Sealey joined the CMF in 1948, was commissioned in 1951 and on retirement in was SO1 Reserves ordnance in Victoria. He resurrected the DRA journal in 1998. He was a Freemason for over 50 years by 2005. *The Australian Reservist*, Commendation, Issue 8, March 2006.

439 DRA Victoria Annual Report 1998/1999, 11 May 1999.

440 Letter, Lieutenant Colonel Sealey to all branch reporters, 15 September 1998.

441 DRA NSW President Report, 29 September 1998.

442 Press Release, Arch Bevis MP, 1 July 1998.

443 Minutes, DRA Victoria, 13 August 1998.

444 Letter, DRA president to State presidents, 3 August 1998.

445 Letter, Major General Glenny to PM, 22 July 1998.

446 *Telegraph*, 'Army Fights Losing War', 9 November 1998.

447 Letter, Major General Glenny to PM, 22 July 1998.

448 ibid.

449 Letter, Lieutenant Colonel Henry to Lieutenant Colonel Hall (DRA NSW), 16 February 1999.

450 Letter, Major General Glenny to State presidents, 14 January 1999.

451 *Sydney Morning Herald*, 21 November 1998.

452 *Canberra Times*, 'Defence Reserves Depleted, says ALP', 10 January 1999.

453 Brigadier McGalliard joined Melbourne University Regiment in 1951 and served in the Reserve for 37 years. He was a civil engineer by profession and was also deeply involved in veterans affairs and welfare. *The Australian Reservist*, Issue 23, October 2014.

454 Facsimile, Brigadier McGalliard to Major General Glenny, 24 March 1999. Victorian branch had its own issues unable to hold an AGM for almost two years due to a lack of quorum to do so despite 150 members in by October 1999. DRA Victoria, AGM Agenda Notes, 31 October 1999.

455 DRA President's Report, 28 March 1999.

456 ibid.

457 *Australian*, 'Falling numbers caught in the Act', 3 May 1999, p4.

458 Speech notes, Bruce Scott MP, at RUSI Canberra, 31 March 1999.

459 For example, see *Hansard*, Representatives, 15 February 1999, p2517. Laurie Ferguson MP (Labor, Reid) was especially effective.

460 Letter, Brigadier McGalliard to PM Howard, as approved 10 May 1999. The DRA AGM and conference was held in May 1999, presumably to compensate against the non-meeting of late 1998.

461 Letter, Brigadier McGalliard to PM Howard, as approved 10 May 1999. The DRA AGM and conference was held in May 1999, presumably to compensate against the non-meeting of late 1998.

462 DRA Victoria Annual Report 1998/1999, 11 May 1999. As well, the DSRC in Victoria was the only one in the country with a DRA member.

463 Minutes, DRA Victoria, 10 June 1999.

464 Minutes, DRA NSW, 6 September 2000.

465 Minutes, DRA NSW, 7 August 2000.

466 Newsletter, DRA Vic, September 2000. The Navy Reserve section of the newsletter was not attributed but was most likely provided by branch vice-president (Navy), Commander M J Hedges RFD.

467 ibid. Newsletter, DRA Vic, September 2000

468 Wilkins, J.M. *op.cit.* p119.

469 Newsletter, DRA Vic, September 2000.

470 Facsimile, Brigadier McGalliard to DRA State presidents, 6 June 1999.

471 *The Australian Reservist*, No3, November 2000.

472 *The Australian Reservist*, No3, November 2000.

473 https://www.aph.gov.au/Parliamentary_Business/Bills_Legislation/bd/bd0001/01bd070

474 Minutes, DRA NSW, 7 August 2000.

475 Minutes, DRA Vic Committee meeting, 13 July 2000.

476 Minutes, DRA Victoria, 14 December 2000.

477 https://www.aph.gov.au/Parliamentary_Business/Bills_Legislation/bd/bd0001/01bd070

478 https://www.aph.gov.au/Parliamentary_Business/Bills_Legislation/bd/bd0001/01bd070

479 Joint Standing Committee on Foreign Affairs, Defence and Trade, 'From Phantom to Force', August 2000, para 7.30, reported in https://www.aph.gov.au/Parliamentary_Business/Bills_Legislation/bd/bd0001/01bd070

480 Ryan, M.J. *op.cit.* p138.

481 News Release, 'Reserve Forces Service and Achievements Noted', Minister for Veterans Affairs and Minister assisting the Minister for Defence, 28 May 2001.

482 Newsletter, DRA Vic, June 2001.

483 *The Australian Reservist*, No4, November 2001.

484 Newsletter, DRA Vic, June 2001.

485 Minutes, DRA Vic, 10 May 2001.

486 *The Australian Reservist*, No4, November 2001.

487 Minutes, DRA AGM, 22 July 2001.

488 Minutes, DRA Vic, 12 July 2001.

489 *The Australian Reservist*, No4, November 2001.

490 ibid. *The Australian Reservist*, No4, November 2001.

491 Tabling Speech, 'A Model for a New Army: Community comments on the 'From Phantom to Force' Parliamentary Report into the Army, R Price, 27 September 2001 as reported in *The Australian Reservist*, No4, November 2001.

492 Minutes, DRA Vic AGM, 18 November 2001.

493 ibid.

494 Reserves formed part of the military part of the Regional Assistance Mission to Solomon Islands (RAMSI), which arrived in the Solomon Islands in July 2003. The 2001 Australian federal election was held in Australia on 10 November with PM Howard's government returned. Senator Robert Hill became Minister for Defence.

495 The Report was compiled by Major General Barry Nunn AO RFD ED (Rtd), former Assistant Chief of the Defence Force (Reserves); Les Cupper, General Manager, Group Human Resources, Commonwealth Bank of Australia; and Peter Kennedy, former Deputy Public Service Commissioner.

496 Letter, Brigadier McGalliard to Brigadier Alkemade, 20 August 2002.

497 ibid.

498 Minutes, DRA Vic, 10 October 2002.

499 CDF Directive No12/2002, CDF to HRP, 19 December 2002.

500 *The Australian Reservist*, No5, November 2002.

501 DRA Vic Annual Report 2001/2002, 10 November 2002. Edwards had served twice in Vietnam in 1968 and 1971 and was heavily wounded in 1971, losing both legs.

502 Minutes, DRA AGM 2001, 15 December 2002. Average attendance was 87.

503 Minutes, DRA AGM 2001, 15 December 2002. This was later changed to be administered under the DRSC.

504 Minutes, DRA Vic, 18 September 2003.

505 *The Australian Reservist*, No 5, November 2002.

506 DRA Vic Annual Report 2001/2002, 10 November 2002.

507 Email, Lieutenant Colonel George to Brigadier McGalliard, 13 December 2002. Vale was Minister Assisting the Minister for Defence from November 2001 to October 2003 and Minister for Veterans' Affairs from November 2001 to October 2004.

508 Minutes, DRA Vic, 16 January 2003.

509 Newsletter, DRA Vic, September 2000.

510 *The Australian Reservist*, No6, November 2003

511 ibid.

512 DRA Conference Q & A paper, 27 August 2005. Flawith transferred to the Reserve on retirement from the Regular Army in 1993. In 2009, he joined the Department of Defence as Director Reserve and Employer Support to head up that area within Cadet, Reserve and Employer Support Division.

513 Minutes, DRA Vic AGM and Meeting with Associations, 23 October 2003. *The Australian Reservist*, No6, November 2003. James was to become a regular presenter and forum coordinator at DRA conferences in the future.

514 DRA Qld Annual Report 2002/2003, August 2003.

515 DRA SA Annual report 2003, 10 August 2003.

516 DRA Vic, 'Meeting with Victorian Liberal Party Defence Reserves Sub-Committee', 15 September 2003.

517 *The Australian Reservist*, No6, November 2003.

518 Gillespie was later to be made a Companion of the order (AC).

519 Minutes, DRA Vic, 7 April 2004.

520 Minutes, DRA Vic, 13 May 2004. Brough was Minister for Employment Services from 2001 to 2004

521 DRA President Annual Report 2004, 25 August 2004.

522 Minutes, DRA Vic, 11 March 2004.

523 Minutes, DRA Vic, 9 September 2004.

524 In 2001, Bailey was made Parliamentary Secretary to the Minister for Defence. In July 2004 she was promoted to Minister for Employment Services and Assistant Minister for Defence.

525 Minutes, DRA, 9 September 2004.

526 Price served as Parliamentary Secretary to the Minister for Defence until March 1993, and also on the Foreign Affairs, Defence and Trade Committee from May 1993 to August 2004.

527 Minutes, DRA Vic AGM and Meeting with Affiliated Organisations, 18 November 2004. Wing Commander Bluck had joined the RAAF Reserve in 1965; he was Chairman of the Victorian DRSC for 14 years to 2007 along with a further three years as DRSC National Executive to 2010. He retired in 2014 after 48 years of service, including 35 years as honorary ADC to eight successive Governors of Victoria. *The Australian Reservist*, Issue 22, April 2014.

528 *The Australian Reservist*, Issue 7, August 2005.

529 Minutes, DRA Vic AGM and Meeting with Affiliated Organisations, 18 November 2004.

530 DRA President Annual Report 2004, 25 August 2004.

531 ibid.

532 DRA President Annual Report 2004, 25 August 2004.

533 *Australian Army Journal*, Vol II, No 1, October 2004.

534 Minutes, DRA Vic, 9 December 2004.

535 Colonel Sandow, comments on article draft, c. February 2005.

536 Email, Major General Glenny to DRA Vic, 30 January 2005.

537 *Defender*, Autumn 2005, pp25-30. Later republished in *The Australian Reservist*, Issue 7, August 2007.

538 ibid.

539 DRA Paper, Conditions of Service, 3 December 2004.

540 Draft Letter to the Editor, Lieutenant Colonel George, 26 February 2006.

541 Directive by CDF, General Cosgrove, 14 May 2005

542 Report on Reserve Remuneration Review, Major General Garde, August 2006.

543 Letter, Bill Shorten to Major General Garde, 18 April 2006.

544 Draft Letter to the Editor, Lieutenant Colonel George, 23 February 2006.

545 Minutes, DRA Vic, 12 May 2005.

546 ibid. By August the Recruiting draft was into its fifth iteration.

547 Lieutenant Colonel Howells, who had developed the Reserve Forces Day website. Minutes, DRA Vic, 14 April 2005.

548 DRA Conference 2005, Theme Introduction, Major General Barry, 27 August 2005.

549 Newsletter, DRA Vic, December 2005.

550 Gambaro was appointed Parliamentary Secretary to the Minister for Defence in July 2004.

551 *Australian Reservist*, No 8, March 2006.

552 DRA Conference Synopsis, Major Rainford, c. September 2005.

553 *Australian Financial Review*, under cover fax Brigadier McGalliard to Major General Glenny, 24 July 2005.

554 Minutes, DRA AGM 2005, 28 August 2005.

555 ADA Defence Brief 123, August-September 2006.

556 Minutes, DRA Victoria AGM, 17 November 2005.

557 Letter to Editor, 'Support defence forces', L. Slocombe, *The Advertiser*, 5 October 2005 and Letter to editor, 'We should back our reservists', Lieutenant Colonel Alexander, *The Advertiser*, 11 October 2005.

558 Minutes, DRA Victoria AGM, 17 November 2005.

559 Executive Circular No 1, DRA President Major General Barry, 31 October 2005.

560 Minutes, DRA Vic AGM, 17 November 2005.

561 Chief of Army, HNA Roadshow presentation, March 2006. The 'R, T & S' roles were given to regional Reserve units to implement in late 2005. Melick retired in 2018 as a major general, when he held the ADF's senior Reserve role responsible for reserves and cadets. A barrister, in 2019 he became national president of the RSL. In the business sector, he was well known in the Tasmanian wine industry as a producer and grower and held several senior civil appointments in Tasmania.

562 Executive Circular No 4, DRA President Major General Barry, 21 May 2006.

563 Executive Circular No 4, DRA President Major General Barry, 21 May 2006.

564 Email, Brigadier McGalliard to Major General Garde, and reply, 13 March 2006.

565 David Binning, 'Special Reserve to be uncorked at the double', *Australian Financial Review*, 22 June 2006.

566 Steve Lewis, 'Army's pay rise won't be enough', *The Weekend Australian*, 25 March 2006.

567 Letter, Major General Barry to Lieutenant General Leahy, 19 August 2006.

568 Letter, Major General Barry to Dr Nelson 30 August 2006 and reply, 13 October 2006.

569 Letter, Invitation to DRA Conference Speakers, Major General Barry, 24 August 2006.

570 The initiative for the new approach to conference organisation was led by Major General Barry's 'task force' of Lieutenant Colonel George, Brigadier McGalliard and Major General Garde.

571 Preparations included receiving helpful comments and suggestions by Neil James, ADA executive director.

572 DRA Conference 2006 draft programme, V3, 30 May 2006.

573 Report on the DRA 2006 Annual Conference, Lieutenant Colonel George, October 2006.

574 *The Australian Reservist*, Issue 9, February 2007. Major General Irving, who had retired from service

but who was Executive Director HR at the NSW Department of Correctional Services at the time, would later become president of the DRA and be awarded the Public Service Medal.

575 Newsletter, DRA Vic, December 2006-January 2007.

576 Report on the DRA 2006 Annual Conference, Lieutenant Colonel George, October 2006.

577 DRA SA branch Report 2006, 20 August 2006.

578 DRA Victoria Annual Report 2005/2006, 27 August 2006.

579 ibid.

580 Minutes, DRA AGM 2006, 27 August 2006.

581 ibid.

582 DRA press release, 'A Seachange for ADF Reservists', 1 October 2006.

583 ibid.

584 DRA press release, 'High Readiness Reserves Initiative is Good but not good enough', June 2006.

585 https://en.wikipedia.org/wiki/Australian_Army_Reserve

586 https://en.wikipedia.org/wiki/Australian_Army_Reserve

587 Newsletter, DRA Vic, December 2006-January 2007.

588 'Report on the Future of the DRA', Major General Barry, October 2006.

589 ibid.

590 President's Report, 2007 DRA AGM, 30 September 2007.

591 Executive Circular No 10, DRA President Major General Barry, 30 September 2007.

592 President's Report, 2007 DRA AGM, 30 September 2007.

593 Minutes, DRA AGM, 7 October 2007.

594 DRA WA branch Report 2005/2006, 27 August 2005.

595 Executive Circular No 8, DRA President Major General Barry, 31 March 2007.

596 DRA Submission, Military Superannuation Review, cMarch 2007.

597 President's Report, 2007 DRA AGM, 30 September 2007.

598 Executive Circular No 10, DRA President Major General Barry, 30 September 2007.

599 Obituary, Major General Cullen, by Major General Glenny, *The Australian Reservist*, Issue 10, February 2008.

600 *The Australian Reservist*, Issue 10, February 2008.

601 ibid. *The Australian Reservist*, Issue 10, February 2008.

602 ibid.

603 ibid.

604 Executive Circular No 10, DRA President Major General Barry, 30 September 2007.

605 President's Report, 2007 DRA AGM, 30 September 2007.

606 The Association later gained $5000 sponsorship for *The Australian Reservist*. Minutes, DRA AGM,

7 October 2007.

607 DRA Queensland branch Report, 5 October 2007.

608 DRA WA branch Report, 7 October 2007.

609 DRA NSW branch Report 2006/2007, October 2007.

610 ibid. The same comment had been made in the annual branch report a year earlier. Annual Report 2005/2006, DRA NSW, cJuly 2006.

611 ibid.

612 A periodic DRA policy review and update was underway at the time, led by a small group which Major General Barry called his 'task force', including national secretary Lieutenant Colonel George, Victorian president Brigadier McGalliard and Major General Garde.

613 DRA Victoria branch Report 2006/2007, 7 October 2007.

614 Fitzgibbon was succeeded as Minister for Defence in June 2009 by Senator John Faulkner (Labor/NSW).

615 DRA National President's Report, Major General Barry, 17 August 2008.

616 DRA Victoria, Annual Report 2007-2008, August 2008.

617 *The Australian Reservist*, Issue 10, February 2008.

618 National President's Report, Major General Barry, 17 August 2008.

619 Minutes, DRA Vic, 12 February 2009.

620 Minutes, DRA NSW, 18 June 2001 and 27 October 2008. Queensland reported 59 members as of July 2007, in a very brief report to the AGM. Annual Report, DRA Queensland, 20 August 2008.

621 https://en.wikipedia.org/wiki/Australian_Army_Reserve

622 Published as 'Stepping Up: Part Time Forces and ADF Capability', in *Strategic Insights*, No44, ASPI, November 2008.

623 *The Australian Reservist*, Issue12, February 2009.

624 Newsletter, DRA Vic, March 2009.

625 https://en.wikipedia.org/wiki/Defending_Australia_in_the_Asia_Pacific_Century:_Force_2030

626 Mark Dodd, 'Army Reserve weak link in defence plans', *The Weekend Australian*, 9-10 May 2009. Minutes, DRA Vic, 14 May 2009.

627 *The Australian Reservist*, Issue 13, September 2009.

628 ibid.

629 DRA Annual Report 2009, Major General Barry, 20 August 2009.

630 Annual Report 2008/2009, DRA Vic, July 2009

631 Annual Report 2008/2009, DRA NSW, cAugust 2009.

632 Annual Report 2008/2009, DRA SA, July 2009.

633 Annual Report 2008/2009, DRA Tasmania, 15 July 2009

634 Annual Report 2008/2009, DRA WA, 21 July 2009.

635 https://en.wikipedia.org/wiki/Australian_Army_Reserve. There was some doubt that Defence knew what the exact numbers were, especially in the Standby Reserve. Even Active Reserve numbers seemed inflated.

636 Newsletter, DRA Vic, March 2010

637 ibid.

638 Newsletter, DRA Vic, March 2010

639 *The Australian Reservist*, Issue 14, March 2010.

640 Newsletter, DRA Vic, July 2010.

641 'The Army Reserve—Back to the Stone Age', *News Weekly*, 26 June 2010, npn.

642 Report, DRA NSW 2009/2010, cAugust 2010.

643 ibid.

644 DRA Discussion Paper, DRA Conference 2010, 11 April 2010. In the 21 August election the Gillard Labor Government was returned, but as a minority government.

645 DRA Conference Report, 'Reserve Futures–Possible Concepts for the Force', 21 August 2010.

646 *The Australian Reservist*, Issue 15, October 2010.

647 ibid.

648 Minutes, DRA Vic committee meeting, 9 December 2010.

649 *Defender,* Vol.II, No1, Spring, 1984.

650 *The Australian Reservist*, Issue 16, April 2011.

651 DRA's Victoria branch noted at its AGM in September 2010 that: 'There is a contradiction between the [DWP] which expands the services and the [SRP] which contracts the Services: both documents are authoritative.' Minutes, DRA Victoria AGM and meeting with Associations, 20 October 2010.

652 https://en.wikipedia.org/wiki/Plan_Beersheba

653 *The Australian Reservist*, Issue 16, April 2011.

654 *The Australian Reservist*, Issue 16, April 2011.

655 Feeney resigned in August 2013 and was subsequently re-elected to the House of Representatives that year (Labor/Batman). He served as Shadow Assistant Minister for Defence from October 2013 to July 2016; Shadow Minister for Veterans' Affairs from June 2014 to July 2016; and Shadow Minister for the Centenary of ANZAC from June 2014 to July 2016. He also served on the Joint Foreign Affairs, Defence and Trade Committee almost continuously from December 2013 to February 2018.

656 *The Australian Reservist*, Issue 17, September 2011.

657 Later Lieutenant General Caligari, DSC AO.

658 *The Australian Reservist*, Issue 17, September 2011.

659 Letter, Lieutenant General Morrison to Major General Barry, 8 August 2011.

660 MONUR had started as a detachment of MUR in 1966 and became its own regiment in 1970. It was finally subsumed as a company of MUR in 2013.

661 *The Australian Reservist,* Issue 18, April 2012.

662 ibid.

663 ibid.

664 A former Regular Army officer, Robert served as Shadow Parliamentary for Defence from December 2009 and as Shadow Minister for Defence Science, Technology and Personnel. After the 2013 federal election Robert was appointed Assistant Minister for Defence in the Abbott Government. After the change of prime minister in September 2015, he was appointed to Minister for Veterans' Affairs, Minister for Human Services and Minister Assisting the Prime Minister for the Centenary of ANZAC.

665 Phillips left the Permanent Navy in 1985 and joined the Naval Reserve in 1992.

666 Interview, Captain Lukaitis with author, 10 October 2020.

667 *The Australian Reservist*, Issue 17, September 2011.

668 *The Australian Reservist*, Issue 18, April 2012.

669 DRA President's Report, DRA AGM 2011.

670 Townsend was appointed Regimental Colonel of the Royal Tasmania Regiment in 2010. He was credited with the success of the Tasman Scheme.

671 Smith was Minister for Defence from 2010 to 2013.

672 *The Australian Reservist*, Issue 18, April 2012.

673 Minutes, DRA Victoria AGM and Meeting with Associations, 26 September 2012.

674 Letter, Major General Caligari to Major General Barry, 9 December 2011.

675 Letter, Major General Sengelman to Major General Barry, 18 April 2012.

676 *The Australian Reservist*, Issue 18, April 2012

677 *The Australian Reservist*, Issue 19, October 2012.

678 *The Australian Reservist*, Issue 19, October 2012.

679 ibid.

680 Letter, Colonel Carey to Stephen Smith MP 10 September 2012 and email, Major General Barry to Colonel Carey, 27 September 2012. DRA president, Major General Barry, delivered a copy of the letter to Senator Feeney on 28 September.

681 Plan SUAKIN generated a whole new world of acronyms.

682 *The Australian Reservist*, Issue 19, October 2012.

683 *The Australian Reservist*, Issue 19, October 2012.

684 Email, Major General Barry to Lieutenant Colonel Ian George, 24 August 2012.

685 ibid.

686 Griggs who later became VCDF, was a former Skyhawk pilot.

687 *The Australian Reservist*, Issue 19, October 2012.

688 Email, Lieutenant Colonel George to Major General Barry 24 August 2012.

689 DRA Queensland, NSW, Victoria Annual Reports 2012.

690 DRA Tasmania Annual Report 2012.

691 DRA SA and WA Annual Reports 2012.

692 As RFSU employment formed part of Border protection there was a case for harmonised conditions of service for all personnel, Reservist or otherwise. DRA President's Paper 'Key issues and Concerns', 8 April 2013.

693 *The Australian Reservist*, Issue 20, April 2013.

694 ibid.

695 *The Australian Reservist*, Issue 20, April 2013.

696 DRA President Paper, 'Key Issues and Concerns, 8 April 2013.

697 https://www.aph.gov.au/About_Parliament/Parliamentary_Departments/ Parliamentary Library/ pubs/ rp/rp1516/ DefendAust/2013

698 DRA President Executive Circulars No1-25, 31 October 2005-14 August 2013.

699 Major General Spence was a former professor of Ancient History at University of New England; he had also been a CO of the 4/19 Hunter River Lancers.

700 *The Australian Reservist*, Issue 21, October 2013.

701 Minutes, DRA AGM, 19 August 2013.

702 DRA President Paper, 'Key Issues and Concerns', 8 April 2013.

703 DRA Victoria Annual Report 2012/13, 7 August 2013.

704 https://www.aph.gov.au/About_Parliament/Parliamentary_Departments/Parliamentary Library/ pubs/rp/rp1516/DefendAust/2013

705 *The Australian Reservist*, Issue 23, October 2014.

706 *The Australian Reservist*, Issue 22, April 2014.

707 ibid.

708 ibid.

709 *The Australian Reservist*, Issue 23, October 2014.

710 *The Australian Reservist*, Issue 23, October 2014.

711 *The Australian Reservist*, Issue 23, October 2014.

712 Minutes, DRA AGM 2014.

713 DRA SA Annual Report 2014.

714 DRA Tasmania Annual Report 2014. Colonel Carey was also Chief Commissioner of the

Workers Rehabilitation and Compensation Tribunal in Tasmania and Chairperson of the Health Practitioners Council as well as Honorary Colonel for the Royal Australian Artillery in Tasmania.

715 DRA WA Annual Report 2014.

716 *The Australian Reservist*, Issue 23, October 2014.

717 https://www.globalfirepower.com/country-military-strength-detail.asp?country_id=australia

718 Atkins, A. 'The Case for A Big Army Reserve', *The Australian Reservist*, Issue 24, April 2015, p28.

719 *The Australian Reservist*, Issue 24, April 2015.

720 ibid.

721 https://www.aph.gov.au/About_Parliament/Parliamentary_Departments/ Parliamentary_Library/ pubs/rp/rp1516/DefendAust/2013

722 A. Hawke and R. Smith, *Australian Defence Force Posture Review*, Department of Defence, 30 March 2012 and https://www.defence.gov.au/publications/reviews/firstprinciples/. A declassified version of the Australian Defence Force Posture Review's final report, which included the estate review, was released by the Minister for Defence Stephen Smith on 3 May 2012.

723 *The Australian Reservist*, Issue 26, April 2016.

724 Email, Colonel Cotton to Major Generals Irving and Barry, 2 February 2016.

725 *The Australian Reservist*, Issue 26, April 2016.

726 *The Australian Reservist*, Issue 25, October 2015. Rear Admiral Firman was later appointed Chief Health Officer for the Department of Veterans Affairs (2019). Rear Admiral Slattery was a Judge of the Supreme Court of NSW. Major General McLachlan was later Commander Land Forces and AO.

727 *The Australian Reservist*, Issue 25, October 2015.

728 *The Australian Reservist*, Issue 25, October 2015.

729 *The Australian Reservist*, Issue 26, April 2016.

730 *The Australian Reservist*, Issue 26, April 2016.

731 *Lancer's Despatch*, No35, August 2018.

732 *The Australian Reservist*, Issue 26, April 2016. ForceNet is a secure ePortal residing outside the Defence Restricted Network enabling communication, coordination, and work force assurance.

733 Minutes, DRA Victoria, 11 February 2016.

734 *The Australian Reservist*, Issue 26, April 2016.

735 *The Australian Reservist*, Issue 27, October 2016.

736 ibid.

737 *The Australian Reservist*, Issue 27, October 2016.

738 Minutes, DRA AGM 2016.

739 ibid. ESORT was the main forum of advice to the Minister of Veterans Affairs, including dialogue

between the Military Rehabilitation Compensation Commission, Department of Foreign Affairs and leadership of the ex-service organisations and Defence communities. The NCF comprised the forums ESORT, National Aged and Community Care Forum (NACCF), Operational Working Party (OWP), Younger Veterans—Contemporary Needs Forum (YVF) and Council for Women and Families United by Defence Service.

740 *The Australian Reservist,* Issue 28, April 2017.

741 ibid.

742 Pay rates were based on the Regular components and used a divisor of 365 days, whereas most organisations used a divisor of 220-250 days. Taxation of Reserve pay would require the introduction of a new and more equitable divisor for calculating the daily rate of pay for reservists.

743 *The Australian Reservist,* Issue 28, April 2017.

744 ibid.

745 ibid. For example, Navy Dive Teams in SA, Victoria, Tasmania and Cairns/Darwin were disbanded. The author's daughter, a Reserve diver, was one of those affected. Captain Lukaitis described the disbandment as a 'very, very, long sad story that goes to the heart of the misuse of the concept of integration within Navy to achieve the outcome of disintegration'. Email, Captain Lukaitis to Major General Barry 9 August 2017.

746 Email Captain Kresse to Captain Lukaitis, 27 June 2017.

747 *The Australian Reservist*, Issue 29, October 2017.

748 ibid.

749 ibid.

750 *The Australian Reservist*, 30, April 2018. The Senate Inquiry was set up in September 2016 and reported to Parliament in August 2017: 'The Constant Battle: Suicide by Veterans'.

751 *The Australian Reservist*, Issue 29, October 2017

752 President's Report 2017 AGM.

753 ibid.

754 Branch reports 2017 Victoria and WA; DRA Treasurer's report 2017.

755 Minutes, DRA AGM 2017; President's Report 2017 AGM.

756 DRA SA Annual Report 2017.

757 *The Australian Reservist*, Issue 30, April 2018.

758 ibid.

759 *The Australian Reservist*, Issue 31, October 2018.

760 *The Australian Reservist*, Issue 31, October 2018.

761 ibid. The RANR post nominal was made redundant in early 2019 to promote a one Navy team culture.

762 *The Australian Reservist*, Issue 32, April 2019.

763 *The Australian Reservist*, Issue 32, April 2019.

764 ibid.

765 DRA Queensland Annual Report 2018.

766 *The Australian Reservist*, Issue 32, April 2019.

767 Governor de Jersey had served with the Reserve between 1966-1971 while at the University of Queensland.

768 Following a review of ADF Cadets, a new HQ ADF Cadets was established in February 2017 and was embedded in CRESD, with HCRESD becoming Commander ADF Cadets.

769 *The Australian Reservist*, Issue 33, November 2019.

770 ibid.

771 President's Report 2019 AGM, DRA Archives. Letter, Major General Irving to Jane McAloon 26 May 2019.

772 DRA President Report, 2019. One outcome of the review was the establishment of the Defence Employer Partnering Network, to support recruitment, retention and utilisation of Reservists to meet capability needs by providing expertise and perspectives on industry and employment matters to Defence.

773 Minutes, DRA AGM, 26 August 2019.

774 Email, Major General Garde to author, 2 November 2020.

775 DRA Victoria Report 2018/2019.

776 ABC News, 25 September 2019.

777 DRA President Report, 2020.

778 Defence Legislation Amendment (Enhancement of Defence Force Response to Emergencies) Bill 2020 Submission 12, Major General Irving, 15 October 2020.

779 ibid.

780 DRA President Report, 2020.

781 DRA President Report, 2020.

782 ibid. The AGM also considered a new idea for a Vice president Legal.

783 Chris Kenny, *Weekend Australian*, 14-15 November 2020, p18.

784 Email, Major General Garde to author, 2 November 2020.

785 ABC News, 30 June 2020.

786 McCarthy, D. *The Once and Future Army, op.cit.* p248.

Index

A

Abbott, Albert, F 28
Abbott, Anthony ('Tony') J, 192, 194
Abigail, Peter J, 158, 163
Afghanistan, 138, 146, 170, 171, 183, 194, 210
 –Karin Towt, 183
Air Force, 18, 19, 115, 169, 175, 186, 188
Alexander, Murray E, 87, 153, 173
Alkemade, Peter D 139, 190
Alliance of Defence Service Organisations (ADSO), 211, 212
Amies, Jack L 2
Amos, R S Philip ('Phil'), 44, 50, 56, 57, 62-64, 67, 74, 77-79, 81-82, 84-89
Andrews, Kevin J 194, 198
Annett, Michael ('Mike') H, 210
Arbuckle, Trevor J, 56, 66
Armed Forces Federation of Australia (AFFA), 51, 165
Army into the Twentieth Century (A21), 106-107, 111-124
Army Reserve Association (AResA) 58, 61-63, 66-69, 71-74, 77, 80-83, 91-93
Army Reserve Review Committee 56, 64
Army Wide Administrative Efficiency Campaign (AWAEC), 63
Army Reserve (ARES)
 –1 Commando Regiment, 146, 170-171, 175
 –1/15 Royal NSW Lancers, 201, 214
 –2 Division 28, 51, 71, 81, 101, 112, 115, 143, 196
 –2 Military District (2 MD), 39
 –2 Training Group, 22
 –3 Division, 2, 11, 17, 63, 84, 85, 88, ,
 –3 Division Field Force Group, 11
 –4 Brigade, 71, 112, 124
 –5 Brigade, 69, 71
 –5 Independent Rifle Company, 33
 –13 Brigade, 166
 –13 Combat Service Support Battalion 217
 –Far North Queensland Regiment, 180
 –1 Battalion, FNQR, 180
 –General Reserve (GRES), 88-89, 101-102, 106-107, 114, 121, 157
 –Melbourne University Regiment (MUR), 8, 182
 –Monash University Regiment (MONUR), 182-183, 186

–Ready Reaction Force (RRF), 9
–Ready Reserve (RRES), 87, 88-89, 102, 104, 106-107, 109, 112, 114, 116, 133, 156-157
–Reserve Staff & Command College, 22, 33,
–Royal New South Wales Regiment (RNSWR), 24
–3 RNSW Training Depot, 15
–4 Battalion RNSWR, 39
–Royal South Australian Regiment (RSAR)
–10/27 Battalion, RSAR, 164
–Royal Victorian Regiment (RVR)
–2 Battalion, RVR, 22
–5/6 Battalion, RVR, 119
–8/7 Battalion, RVR, 125
–Sydney University Regiment, 2
Standby Reserve, 138, 150, 166, 174, 197, 202
Arnold, Michael ('Mike'), 204
Arundel, Bartholomew, 219
Asia Pacific Economic Cooperation (APEC), 164
Association of Australian Reserve Forces, 73
Assistant Chief (of the General Staff) Reserves–Army (ACResA), 66, 86
Assistant Chief of Defence Force–Reserves (ACDF-R) 98, 102, 124, 143, 146, 150, 164
Atkinson, Robert ('Bob') N, 208
Auditor-General (A-G), 84, 172
Austin, Tony K, 162
Australian, 65, 112, 122
–*Weekend Australian*, 172
Australian Cadet Corps, 57, 91
–Regional Cadet Units (RCUs), 91

Australian Defence 1994, 103
Australian Defence Association (ADA), 22, 33, 51, 105, 112
Australian Defence Force (ADF), 34, 59, 66, 108, 111,
–Total Workforce Model (TWM), 205, 207, 210
Australian Financial Review, 132, 153, 156
Australian Imperial Force (AIF), 3
Australian National Audit Office (ANAO), 84
Australian Regular Army (ARA), 9, 19,
–Army Headquarters (HQ), 89
–Eastern Command (E Comd), 1, 6
–Field Force Command, 55, 57
–Kapooka, 142
–Keswick Barracks, 181, 188, 193
–Land Force, 69, 70
–Randwick Barracks, 87, 131, 157, 193
–Royal Australian Armoured Corps (RAAC), 201, 202
–Royal Australian Regiment (RAR), 4
–Simpson Barracks, Watsonia, 142
–Southern Command (S Comd), 2, 18
–Victoria Barracks (Sydney), 28, 56, 164
–Western Command (W Comd), 6
Australian Security Intelligence Organisation (ASIO), 80
Australian Strategic Policy Institute (ASPI), 158, 163, 171, 187-188, 196
Australian Veterans and Defence Services Council (AVADSC), 51, 52, 153, 165

B

Bailey, Frances ('Fran') E, 146
Baker, John S, 107, 115-116
Barnard, Lance H, 9, 14
Barr, Alan, 100

Barr, Ian ('Sam') F, 56
Barry, James ('Jim') E, 160-162, 165-166, 169-171, 173-175, 180, 182, 184-188, 192-193
Baume, Michael E, 70
Bedggood, Elizabeth D, 141, 162, 184
Behm, Anthony ('Tony') P 117
Bennett, Jennifer 181
Bennett, Phillip H 35-37, 47, 59
Bevis, Archibald ('Arch') R 107, 123
Bidgood, Anthony ('Tony') C, 56
Billson, Bruce F, 163
Bilney, Gordon 84, 90
Birmingham, Mathew, 151
Bishop, Bronwyn K, 118
Bishop, John A 2
Bishop, T Mark ('Mark'), 151
Bluck, Richard J, 146
Bougainville, 138
Brereton, Paul L G, 183, 188
Bridie, Phillip K H, 214
Broadbent, John R 74, 84, 119
Broadhead, Kenneth ('Ken') J, 57, 65
Brodtmann, Gai M, 211
Brogan, Mervyn F, 13, 29, 31
Brough, Malcolm ('Mal') T 145
Brown, Clay, 215
Brown, John J, 56
Browne, Anthony ('Tony') C, 2, 5, 13
Byrne, R J (fnu) 66, 78
Bullard, David E F, 15, 17, 25, 28, 44, 52, 56, 65
Bushmaster, 157, 188, 190

C

Cadet, Reserve and Employer Support Division (CRESD), 180
Caligari, John G 172, 182, 185, 186

Callinan, Bernard J, 40
Call-Out Legislation, 56, 63, 79
Camp Bradman, 161
Campbell, Angus J, 204-205, 207
Campbell, Kathryn J, 213, 218
Carey, Stephen ('Steve'), 188, 196
Cawthorn, Gary, 27
CGS Advisory Committee (CGSAC), 61
Chester, Darren J, 200, 209, 212, 218
Chief of Air Staff (CAS), 115
Chief of Defence Force Staff (CDFS), 15, 38, 65, 98
Chief of the General Staff (CGS), 6-7, 12, 15, 25, 35-38, 50, 53, 56, 61, 64-65, 69, 72, 84-87, 98, 103, 115-119
Chief of Reserves (CRES), 28, 52-53, 57, 65, 69
Choy, Darryl C L, 136
Citizen Military Forces Association (CMFA), 1-23, 25-29, 32-35, 37-45, 48-59
Citizen Military Forces (CMF), 1-14, 17-25, 29, 35-37, 39, 42, 48, 50
Citizen Soldier 2, 4-5, 9-10, 12-13, 23, 31, 35, 38, 40, 42, 48, 87
Civil Military Cooperation (CIMIC), 143
Clendinnen, Brian D, 56, 64-65, 68, ,
Clunies-Ross, Adrian, 56
Coates, H John ('John'), 104, 112
Coldham, Peter A, 22
—Coldham Committee, 22, 39-40, 42-43, 45
Committee for Employer Support of Reserve Forces (CESRF), 19, 24, 35-36, 38, 143
Committee of Employer Support Army Reserve (CESAR), 15, 19
Committee of Reference for Defence Force Pay, 22, 39

Common Induction Training (CIT), 122, 124-125, 128, 144, 150
Common Recruit Training (CRT), 141, 144
Compensation (Commonwealth Employees) Act 1971, 23
Connolly, James M, 121
Continuous Full Time Service (CFTS) 215-216
Cook, Bruce, 170, 172
Cook, Richard ('Dick') G 196
Cooke, Kevin G 32, 50, 53, 56, 63-64, 68, 74, 78, 89, 107, 143, 145, 152
Cory, Gilbert E, 2
Cosgrove, Peter J, 136, 139, 152
Cossum, Elizabeth ('Liz'), 218
Cotton, Jennifer ('Jenny'), 184, 196, 214
Courier Mail, 160
Crackett, John, 199
Crawford, T J (fnu), 13, 32
Crews, William ('Bill') J, 162
Crompvoets, Samantha, 188
Crook, J R W (fnu), 2, 31
Crosby, Jed, 203
Cross, Manfred D, 66-67, 79
Cross Committee 72, 75, 77, 80
Cullen, Paul A, 1-7, 9-11, 15-16, 18-19, 21, 27, 29, 31, 37, 39, 45, 47-48, 58, 62, 65, 67, 69-72, 74, 77, 80, 84, 86, 88, 92, 99, 101, 103, 111, 113, 116, 119-122, 124, 131, 145, 163, 164
Cuneen, Chris. 205
Cutler, Arthur Roden, 28, 119, 123

D

Daly, Thomas, 13
Dance, Michael ('Mike'), 163
Dart, John R, 2-3, 13, 17, 22, 31, 38, 64, 67
Davies, Andrew, 171, 196

Deane, William P, 123
Defence
–Defence 2000: Our Future Defence Force, 132
–Defence Consultative Committee (DCC), 131, 132
–Defence Efficiency Review (DER), 118, 121, 126, 133
–Defence Force Welfare Association (DFWA), 165, 170-171, 211
–Defence Forces Advancement Group, 18
–Defence Forces Remuneration Tribunal (DFRT), 45, 48, 56-57, 65-66, 100
–Defence Forces Retirement and Death Benefits (DFRDB), 53, 81
–Defence Home Ownership Assistance Scheme (DHOAS), 191-192, 195-198, 208
–*Defence Legislation Amendment (Enhancement of the Reserves and Modernisation) Bill 2000*, 131, 216
–*Defence of Australia 87 (DOA 87)*, 88
–*Defence Re-establishment Act 1967*, 131
–Defence Reform Program (DRP), 120-121
–Defence Reserves Association (DRA), 29, 87, 90, 92-94, 97-102, 104-109, 111-133, 135-153, 155-214, 216-225
–*Defence Reserve Service (Protection) Act 2001*, 135, 163, 170, 173, 191, 195, 197, 201, 207, 210
–Defence Reserves Support Committee (DRSC), 102, 143, 146, 152, 165, 169, 171, 195-197, 204, 207, 213-214
–Defence Reserves Support Day (DRSD), 173
–Defence Service Homes, 16
–Defence Support Organisations Secretariat (DSOS), 45

–Defence White Paper (DWP)
–Defence of Australia 1987, 61
–Defender, 148, 178
–Defending Australia in the Asia Pacific Century: Force 2030, 172
de Jersey, Paul, 213
de Laat, F Karel, 140, 143
Delahunty, Terence ('Terry') C, 177
From Phantom to Force, 130, 133
Department of Veterans Affairs (DVA), 122, 176, 197, 205, 207, 209, 211-212, 218-219
Deputy Chief Army (DCA), 122
Deputy Chief of Reserves (DCRES), 57
Dibb, Paul, 45
Dibb, Report, 47, 49, 53-55, 58, 61, 63-64, 67, 74
Director of Reserves–Air Force (DRES-AF), 100
Director General Naval Reserves (DGNR), 183
Director General Australian Navy Cadets and Reserves (DGANCR) 206, 218
Director Naval Reserves (DNR), 117
Dodd, Mark, 172
Donovan, Patrick R, 78, 99
Dougherty, Ivan, 3, 13, 31
Duffy, Michael J, 50
Dunlop, David J, 143
Dunbar, Wayne L, 143
Dunn, Jeff, 13, 15, 32, 58
Dunn, Peter, 106, 116
Dunne, Kathryn, 205
Durack, Peter, 84

E

Eason, Richard ('Dick') T, 2, 15
Eastick, Thomas, 3, 13

East Timor, 130-131, 136-138, 142-143, 152, 165, 223
Eather, Kenneth W, 3, 78
Edwards, Alan, 32, 35-36, 45
Edwards, Graham J, 140
Efficiency Decoration (ED), 11, 17, 19-20
Efficiency Grant, 9
Elkington, Joanna ('Jo'), 213
Elms, Adam, 203
Elsey, Ranford, 172-173, 177, 193,
Employer Support Payment Scheme (ESPS), 189, 191, 194-195
Ennis, G J (fnu) 85
Ernst & Young, 181, 205
Evans, Bernard, 3
Exercise
    –ANZAC, , 185
    –Kangaroo-89, 73, 79
    –Talisman Sabre, 180
Ex-Service Organisation (ESO), 171, 196, 208, 217, 225
    –Ex-service Organisations Round Table (ESORT), 205, 208-209, 211

F

Faulks, Ronald M, 13, 31
Fawcett, David J, 199
Fay, Rodney G, 66, 74-75, 78, 80, 82, 84, 87-89, 93-94, 97-102
Feeney, David I, 177, 182, 184-185, 189, 191-192, 195, 198-199, 207
Ferguson, Grant I, 204
Ferguson, Laurie D T, 131, 136
Fleeton, Graham, 123, 167, 176
Finane, Michael J, 80
Firman, Jennifer ('Jenny') R. 199, 204
Flawith, Ian B, 143, 162, 173, 193

Force Modernisation Review, 180-181, 223
ForceNet, 202
Force Protection Companies, 157
Force Structure Review (FSR), 87-89, 99, 107, 114, 133
Forde, Leneen, 151
Foster, Hamish, 144
Fox, Natasha A, 218
Fraser, Bryce, 71
Fraser, Colin A E, 18
Fraser, J. Malcolm, 12, 15-17, 22, 25
Frazer-Jans, Judy, 81
From Phantom to Force, 131, 133
Full Time (F/T), 109, 114, 174, 177, 197
Full Time Duty (FTD), 117, 145
Furse, Anthony ('Tony') S, 56

G

Gambaro, Teresa, 150
Garde, Gregory ('Greg') H, 56, 65, 101, 136, 139-140, 143, 147, 149, 152, 156, 163, 165, 168, 171-172, 174-177, 180, 182, 190-191, 195, 200, 210
Garland, Ronald S, 13, 31
George, Ian 104, 174
Gibson, Brian W, 78
Gilbertson, Colin ('Col'), 97, 116, 120-122, 129, 170
Gillespie, Kenneth ('Ken') J, 145, 153, 158, 174
Glenny, Warren E, 7, 10, 22, 26, 29, 31, 38, 78, 101-102, 106-109, 111-117, 119-128, 130-131, 135, 137-138, 140-148, 150, 152, 163, 165, 171, 193, 200
Godfrey, Michael, 119
Golding, Stephen, 103, 117
Goodwin, Luke, 185
Gould, Simon G, 210

Grant, William H ('Mac') 15
Gration, Peter C 47, 50, 52, 61, 62, 69, 71, 79, 98
Greaves, David 213
Greeles, Don 112
Green, Kenneth D, 17
Greet, Neil, 165
Grey, John C 98, 103
Griggs, Raymond ('Ray') J 189, 195
Grimes, Laureen, 214
Guard's Concert, 4

H

Haddad, Peter F, 146
Halverston, Robert G, 49
Hamer, Rupert 28
Hardened and Networked Army (HNA), 155-156, 157, 188
Hardy, Peter A, 105
Harman, Jac, 217
Harradine, R Brian W, , 41, 51
Harris, Graham, 32
Harrison, R I ('Sam'), 56, 59, 63-64, 66-67, 73, 78
Hartley, John C, 122
Hassett, Francis ('Frank') G, 7, 15
 –Hassett Committee, 7
Hawke, Alexander ('Alex') G, 212
Hawke, Robert ('Bob') J L, 38, 41, 42, 49, 61, 71
Haynes, John S, 108, 165
Hayward, Duncan, 218
Head Modernisation and Strategic Planning–Army (HMSP-A), 186, 188, 196, 199
Healy, Robert ('Bob'), 27
Hedges, Malcolm J, 108, 129
Herald-Sun, 90

Herring, Edmund F, 3, 13, 21, 35
Hickling, Francis ('Frank') J, 116
Hicks, Peter, 153
High Readiness Reserve (HRR) 128, 142, 151, 155, 157, 160, 169, 223
Hill, Mark D, 210
Hill, Robert, 78
Hindmarsh, Tony, 210, 218
Hines, Colin Joseph, 12-13, 31
Houston, A G ('Angus'), 150, 152
Howard, John W, 108, 119, 124, 126, 128, 148-149
Hulse, George L, 107, 109
Hume, Robert W G, , 93, 129
Humphries, Gary J J, 177
Hurley, David J, , 177

I

Ikins, Charles G, 177
Imperial Services Club, Sydney, 1, 5, 35, 50
Inactive Reserve, 138
International Stabilisation Force (ISF), 185
Irving, Paul, 140, 169, 171, 180, 183, 190, 194-195, 197, 200-206, 208-209, 211, 213, 216, 218-219

J

James, Neil, 144, 146, 148, 151, 157, 163, 165, 204
James, William B ('Digger'), 101
Jans, Nicholas A, 81, 83, 106, 170, 177
Jennings, Peter 117
Jessup, Christopher N, 56-57
John, Anthony, 146
Joint Standing Committee on Foreign Affairs, Defence and Trade (JSCFADT), 130, 133, 138, 199, 205
Johnson, Joseph ('Joe') V, 61
Jongeneel, Constance ('Connie'), 161
Just, Malcolm E, 13

K

Kafer, Bruce 199, 204, 207, 209
Kashmir
 –Kashmir Line of Control, 161
Katter, C Robert ('Bob') Snr. 34
 –Katter Committee, 34
Keating, Paul J, 45
Keefe, John P, 57, 65
Keldie, John D ('Blue') 70-71, 74, 80, 84, 97
Kelly, Michael ('Mike') J 169, 207
Keys, A G William ('Bill') 45
Khan, Colin N ('Ghengis'), 32
Killen, D James ('Jim'), 13-16, 23
King, Leslie D, 78
Kirk, Matthew, 215
KPMG, 213
Kresse, Frank 206

L

Laidlaw, Douglas ('Doug') W 218
Land Warfare Centre 48, 143
Laube, David 164
Leahy, Peter F, 139, 146, 148, 157, 168
Lee, Arthur J, 28
Legislation Amendment (Enhancement of Defence Force Response to Emergencies) Bill 2020, 215-216, 218
Liberal
 –Defence Committee, 8, 49
Lindsay, Eamon ('Ted') J 80
  Lyndsay Committee, 82, 91
Lindsay, Peter J, 84, 89, 163, 169, 172
Long, Lee T, 201-202
Lowen, Ian H, 13, 17-18, 23-24, 26-27, 29, 32, 34, 40-41, 44-45, 47-59, 63, 68, 85, 91, 162

Lukaitis, Joseph L, 171, 183-184, 189, 193, 199, 202, 206-207
Luttrell, Denis R, 99-100, 102-103

M

Macarthur-Onslow, Denzil, 3, 13, 31
Macdonald, John A. L. ('Sandy'), 158
Macdonald, John Mc L, 23, 26, 27, 31, 35, 38
MacDonald, Arthur L, 15
MacGibbon, David J, 84, 92, 131-132
MacLennan, Lewis, 203
Magee, Owen, 22
Maitland, Gordon L, 6, 22, 24, 26, 32, 35, 38-39, 44, 84, 119, 121
Marles, Richard D, 203, 213
Martin, James E, G ('Sparrow') 78
Martindale (fnu) 80
Mayer-Frisch, Rainer H, 142-143
McAloon, Jane F, 207, 209, 213
McClelland, Robert B 158
McCarthy, Dayton, 1, 7
McCarthy, Dennis M, 177
McDermott, Peter J 170, 173
McDonald, Bruce A 12
McDonald, Stuart M, 2
McGalliard, Andrew ('Andy') J, 127-130, 132, 135, 139, 144, 149, 152-153, 156, 168-170, 176
Mcintosh Committee – also see DER, 118
McLachlan, Fergus ('Gus'), 199
McLachlan, Ian M, 116-119, 124
McCracken, Cameron G 165
McGrath, Brian A 106, 115
McKinna, John G 32, 36
McLeod Donald ('Don') H 17, 26
Mead, Dennis 27
Melick, Aziz ('Greg'), 155, 164, 169, 177, 204
Mercer, Robert ('Bob') D, 32
Millar, Thomas Bruce ('TB'), 8
 –Millar Report, 7-12, 14-16, 1921-22, 43, 55, 63, 68, 72, 92, 107, 133, 188
Military Board
 –CMF Member, 1, 3, 5-7, 12, 18, 86, 163
Military Compensation Scheme (MCS) 78, 80-81, 99, 167
Military Rehabilitation and Compensation Amendment (MRCA), 212
Minchin, Nicholas ('Nick') H 169
Minimum Level of Operational Capability (MLOC), 101
Minister for Defence (MINDEF), 8, 13, 15, 35, 37-38, 41, 43, 49-50, 52, 54, 56, 64-65, 67, 69, 71, 74, 79-80, 82, 84, 87, 90-91, 98-99, 107, 111, 113-114, 116-119, 123-124, 127, 129-131, 137, 140-141, 145-146, 150-151, 155, 157-158, 163, 165, 168-169, 177, 181-183, 185, 188-189, 194-195, 197-200, 203, 207, 209, 212-213, 216, 218
Moffitt, Rowan C, 165
More, John C, 128
Morkham, John, 162
Morrison, David L, 182, 185
Morrison, Scott J, 209, 224
Mott, Ian 74
Murchison, Allan C, 1, 3, 6, 12-13, 15-16, 19, 23, 27, 31, 52, 84, 119
Murray, Kevin R 35, 40, 72, 74, 80, 84

N

National Consultative Framework (NCF), 205
National Service (1964-1972), 4-5, 7, 68, 159
National Servicemen's Association (NSA), 122

Naval & Military Club, Melbourne, 17, 50
Nebenzahl, Brian, 2, 13, 31, 38, 44, 51, 120
Nelson, Brendan J, 155, 157
Neuman, Shayne K, 213, 218
Newman, Jocelyn M, 84
Newman, Kevin, 32
Noble, Roger J, 218
Non-Commissioned Officer (NCO), 9, 14, 25, 49, 51, 69, 83, 98, 125, 145, 173, 179, 223
North-West Mobile Force (NORFORCE), 33
Norris, John G, 40
Nunn, Barry N, 61, 98, 100, 102
   –Nunn, Report, 139, 149
   –Nunn/Sanderson Report, 61, 67-69, 133

O

O'Brien, Michael ('Mike'), 146
O'Connor, Michael J, 105, 112, 136, 140
O'Donnell, Lawrence ('Laurie') G, 55, 69, 72, 77
O'Dempsey, Keith, 66, 78-79
Office of the Chief of Reserves, 40
Officer Cadet Training Units (OCTUs), 103
Operations
   –ANODE, 144, 193
   –ASTUTE, 193
   –BUSHFIRE ASSIST, 218
   –CITADEL, 211
   –COVID-19 ASSIST, 215, 217-219, 224
   – DELUGE, 164
   –SLIPPER, 194
   –TANAGER, 137
Operational Level of Capability (OLOC), 101
Order of Battle (ORBAT), 192

Orme, Craig W 207
Other Ranks (ORs), 189

P

Pacific Defence Reporter, 11
Pagan, John Ernest, 1-2, 13, 31
Part Time (P/T), 109, 197, 205-206
Part Time Duty (PTD), 145
Parsonage, Philip C, , 67
Payne, Marise A, 165, 200
Peacock, Andrew S, 131
   –Peacock Committee, 132, 135, 142
Pennicook, Douglas, 2
Pentropic Division, 7, 11, 116
Phillips, Richard 169, 183-184
Pixley, Neville, 21
Plan BEERSHEBA, 179, 180, 182, 185-187, 191-193, 201, 207, 210, 222
Plan SUAKIN, 179, 182, 192-193, 201, 210, 222
Poke, Francis E, 15, 17, 25-26
Pollard, Bill, 27
Pollard, Reginald ('Reggie') G, 116
Porter, Stephen H, 201
Pratten, Garth, 156
Price, L Roger ('Roger') S, 138, 146
Price, Melissa L 212
Project WELLESLEY, 102, 106, 108, 113, 133

Q

Queen's Jubilee Medal, 52
Quelch, Hugh, 120
Quick, David M, 56
Quirke, Peter J, 149

R

Rainford, Ian S, 151-152
Ray, Robert F, 80, 87, 98

Ready Response Force, 144
Regional Assistance Mission to Solomon Islands (RAMSI), 144
Regional Force Surveillance Unit (RFSU), 191
— North-West Mobile Force (NORFORCE) – see NORFORCE
Reith, Peter K, 104
Reserve Battalion Battle Group, 187
Reserve Forces Day (RFD), 122-123, 126-127, 138-139, 159, 168, 171, 173, 197
Reserve Forces Decoration (RFD), 17, 21
Reserve Forces Support Day (RFSD), 168, 173
Reserve News, 130
Reserve Modernisation Workshop (RMW), 181
Reserve Remuneration Review, 149, 170, 173
Reserve Service Days (RSD), 212, 215
Returned Services League (RSL), 13, 18, 22, 31, 33, 45, 51-52, 94
— Kindred Organisations Committee (KOC), 94, 97, 101
Restructuring the Army (RTA), 122
Review of Defence Force Remuneration – see Nunn Report
Review of Employment Specifications, Trades and Army Reserve Trade Training Requirements (RESTARTTR), 63
Reynolds, Linda K, 212-213
Risson, Robert J H, 3, 40
Robert, Stuart R, 183, 189, 192, 195, 197
Roberts, Francis ('Frank') X, 142
Roberts, John B, 2, 13
Rodgers, Robert ('Bob') P, 204
Rosenfeld, Jeffrey V, 172, 177
Rossi, Keith V, 15, 25

Royal Australian Air Force (RAAF), 22, 25, 116, 129, 203-204
— Butterworth, Malaysia, 131
RAAF News, 115
RAAF Active Reserve, 117
RAAF Reserve 18, 22, 40, 72, 111, 112, 115, 143-144, 166, 168-170, 186, 215
— 21 Squadron, 90
— 22 Squadron, 33
— 25 Squadron, 85
— Air Force Emergency Reserve, 19
— Citizen Air Force (CAF), 18
— Permanent Air Force (PAF), 18, 115
— RAAF Amberley, 106
— RAAF Laverton (see RAAF Williams)
— RAAF Learmonth, 85
— RAAF Point Cook (see RAAF Williams)
— RAAF Williams, 169, 172, 193
Royal Australian Navy (RAN), 3
— Australian Naval Reserve (ANR) 105, 106
— Fleet Base West 192
— HMAS *Stirling* 192-193
Chief of Naval Staff (CNS) 86
— HMAS *Manoora,* 151
— Permanent Naval Force (PNF) 105, 129-130, 136, 150
Royal Australian Navy Reserve (RANR), 21, 27, 32, 37, 40, 93, 98, 100, 108, 130, 162, 169, 172, 184, 199, 202, 204, 206-207, 210
— HMAS *Adroit,* 92
— HMAS *Ardent,* 37
— HMAS *Bayonet,* 34
— HMAS *Lonsdale,* 150
— HMAS *Manoora,* 151

- General Reserve, 106, 172, 183
- Naval Ready Reserve, 106
- Standby Reserve, 106

Royal Military College Duntroon (RMC Duntroon), 70, 103

Royal United Services Institute (RUSI), 86, 119, 128, 165, 208

Regimental Sergeant Major (RSM), 200

Rudd, Kevin M 168, 192

Rule, Peter, 144

Rum Corps, 1, 62

Rwanda, 106, 114

Ryan, Alan, 140, 143, 146, 158

Ryan, Maurie, 133

S

Safety, Rehabilitation & Compensation Act (SRCA), 99

Sanderson, John M, 61, 115-117, 119, 128
- See Nunn/Sanderson Report

Sandow, Donald ('Don') J 44, 66, 90, 98-107, 109, 113-114, 116, 118-119, 120, 131, 148, 162

Sandow, John, 27

Sankey, Catalina, 205

Scherger, Frederick R, 21

Schiller, Carl F, 169, 196, 205

Scott, Bruce C, 127-128, 137

Sealy, Robert, 120, 122-123, 162

Sealey, Beryl, 162

Self-Employed Reservists (SER), 131, 135, 189, 213

Senate
- Standing Committee for Foreign Affairs and Defence, 8, 48

Sengelmann, Jeffrey ('Jeff') J, 186, 188, 196

Service Categories (SERCAT), 186, 198, 213

Service Options (SEROPS), 186

Seselja, Zdenko ('Zed') M, 195

Sharkey, G J, 141, 169

Sharp, Raymond ('Ray'), J, 35, 72-73, 79, 84, 92, 97, 102-106, 111, 116, 119, 121

Sher, Gregory M, 171

Shirtley, Graeme S, 162, 165, 170, 173

Shorten, William ('Bill') R, 150

Sibree, Peter A, 61

Simmons, David W, 71-72

Simpson Barracks, Watsonia, 142

Simpson, Lionel A 13, 15, 32, 41, 51

Simpson, Noel W 3

Slattery, Michael, 218

Slattery, Michael J 199

Smethurst, Neville R, 69-72, 75

Smith, Hugh, 104, 112, 133, 165, 170, 177, 188

Smith, Steven L, 196

Smith, Steven F, 185

Smorgon, Jack, 214

Special Operations Task Group (SOTG), 175

Spence, Iain, 177, 189, 192, 196, 199, 204

Spicer, Ian, 40

Standish, Graeme B, 80, 89
- Standish Report, 93, 99-100, 102

Stanley, Raymond ('Ray') J, 51, 63

Staziker, George S, 57, 65

Stetson, John, 36

Strutt, Horace ('Harry') W, 28

Sydney Morning Herald, 126

T

Tabuai, Karl, 180

TAC5, 48

Tan, Audrey, 161

Tange Review, 122
Tasman Exchange Scheme, 66, 141
Taylor, Kenneth V, 100, 117
Tertiary Education Assistance Scheme (TEAS), 44
The Australian Reservist 87, 122, 130-131, 136-137, 141, 144, 147, 162-163, 169, 171, 175, 179, 181, 184, 191, 194, 197, 204, 206, 217, 219, 224
The Kensingtonian, 20
The Reservist Association (TRA), 73, 79, 92
Thirwell, R S J, (fnu), 99
Thomson, Jessica, 156
Thyer, James ('Jim') H, 2
East Timor/ Timor Leste, 130-131, 136-138, 141-143, 152, 165, 183, 185, 193-194, 211, 223
Total Force 11, 54, 68, 75, 107, 115, 117, 125, 131-132, 145-146, 181-182, 185, 195, 197-199, 213-214, 223-224, 225
Townsend, Dennis, 184, 190
Turnbull, Malcolm B, 203, 209
Turner, Neil, 143
Tyler, Alex, 185

U

United Arab Emirates
 –Camp, Baird, 203
United Services Club, 203
United Services Institute (USI), 47

V

Vale, Danna S, 140-141
Veterans and Defence Service Council (AVADSC), 22, 51, 165
Veterans Entitlement Act (VEA), 81
Vice Chief Defence Force (VCDF), 124, 153, 158, 173, 175, 177, 195
Vickery, Norman A, 5-6, 18, 85

Vietnam Medal, 52
Vietnam War, 4-5, 8, 222
Vincent, Michael, 101
Von Nida Dewes and Sachs Pty Ltd, 74

W

Western Soldier, 66
Wertheimer, Peter W, 169, 185
White, Andrew, 218
Whiting, Kevin D, 32
Whitlam, E Gough, 8, 12
Wiggins, Stephen, 27
Wilkins, John M, 130
Willis, Max F, 68, 71, 124
Wilsonaurie, 34
Wilson, Neil, 136, 140, 173, 175, 180, 191, 193, 196, 210-211, 214
Wilson, Richard G, 146, 150-151, 156, 158
Windeyer, Victor, 3, 13, 31
Wolski, Brett S, 213, 218
Wood, James ('Jim'), 183
Workplace Relations Act, 123
Wright, Norman, 17
Wrigley, Alan K C N, 80, 82
 –Wrigley Report, 81-83, 88
Wyatt, Douglas ('Doug') M, 190

Index

www.ingramcontent.com/pod-product-compliance
Lightning Source LLC
Chambersburg PA
CBHW060920170426
43191CB00024B/2445